Nuclear Reactors

Red Alert is the deliberately urgent title for a new series of Earth Island texts dealing with specific issues on an ever-growing front of environmental concern.

The texts are short, from 15,000 to approximately 25,000 words, the simple assumption being that if the problem is as pertinent as is claimed, and the writer sufficiently sure and lucid in the handling of his material, he and his publisher should be capable of producing a work quickly, which the intelligent layman will read and absorb easily, and in certain ways, act upon.

Though brief, 'Red Alerts' are authoritative, comprehensively annotated when necessary, and include both index and bibliography. In the sense that any book must be of its time and place to be of real value, both their frames of reference and their data are contemporary and apposite. And in the sense that no book can ever be complete, they are calculated to lead both author and reader on to further examination of the subject.

First titles are *Nuclear Reactors* and *Openpit Mining*. Additional titles Scheduled for 1973 are an investigation of The Bread Industry, Foulness, and Transport. Further titles may include The Oil Industry, Plastics Industry, Supersonic Flight, The Drug Industry . . .

Sean F. Gallagher, Series Editor

Nuclear Reactors

Walter C. Patterson

No. 1 in a series of urgent environmental studies

Earth Island

First published in Great Britain 1973
Earth Island Limited
56 Doughty Street
London WC1N 2LS

ISBN 0 85644 016 7

Text set in 11/13 pt. Monotype Baskerville, printed by letterpress, and bound in Great Britain at
The Pitman Press, Bath

To my parents, who didn't worry when I dropped nuclear physics, and to Cleone, who didn't worry when I picked it up again.

Contents

Preface

Nuclear reactors: are they just a fact of life, or a matter of life and death? As reactors spring up everywhere the question is becoming ever more urgent. In books, magazines and newspapers, on television, and in the public forum the debate grows steadily more heated. Alas for clarity; nuclear reactors have always been veiled in an esoteric mystique, making the debate difficult to follow and still more difficult to evaluate. What are they talking about?

But a nuclear reactor is scientifically no more mysterious than a transistor radio. This book is intended to dispel, at least somewhat, the impression that nuclear reactors can only be understood and discussed by the 'experts'. Too many 'experts' have already spoken too much nonsense. In the field of nuclear technology what we now need most are informed and articulate laymen. The 'experts' have been allowed to grind their nuclear axes for too long in privacy. The neck they are risking may be yours.

Walter C. Patterson, 1973

1

WHAT IS A REACTOR?

Atom and nucleus

If you take a pair of metal hemispheres and slam them together very fast, face to face, one of two things may happen. You may get a loud clunk. Or you, the hemispheres, and everything else in the vicinity may be almost instantly vaporised in a burst of incredible heat. If the latter happens, you can be sure that the metal was a particular kind of uranium, not that the confirmation will do you much good.

What has vaporised you is raw energy, released from the innermost structure of the uranium. The energy in the interior of uranium was revealed to the world on 6 August 1945, in the sky above Hiroshima, Japan; never has a new source of energy made such a horrifying début. Yet, paradoxically, the most overpowering energy man has released comes from the tiniest reservoir he has yet learned to tap: the nucleus of an atom.

What is an 'atom'? And what is its 'nucleus'? Take a lump of lead, and cut it into smaller and smaller pieces. When the pieces are so small that your knife is too clumsy, switch to an imaginary knife and keep cutting. Ultimately, the pieces will get so small that if you cut any more you will not get two pieces of lead: the next cut will change the identity of what you are cutting. The smallest piece which is still lead is called an atom of lead.

The word 'atom' means 'indivisible'. You cannot divide an atom of lead and still get lead. But you can divide the atom and get smaller pieces which are no longer lead. If you start to dismantle an atom, the first part you get off is called an 'electron'. Until this stage, you have been able to cut without encountering any conspicuous electrical effects; but the electron is negatively charged, and the remainder of the atom is left positively charged. They have become 'ions'. Each time you remove another electron you leave still more positive charge on the remainder. Since negative and positive charges attract each other, it is harder and harder to pry off successive electrons.

Suppose, however, that you manage to remove *all* the electrons (which for most atoms would, in practice, be very difficult). What you would have left would be the innermost heart of the atom: the nucleus. This is where all the positive charges are. Furthermore, you would now find an enormously increased difficulty in doing any more cutting. Surprisingly enough, although the nucleus contains only positive charges (which of course *repel* each other), its constituent pieces cling together with a loyalty which makes the outer electrons look frankly promiscuous.

There used to be much popular talk of 'splitting the atom'; but for scientists the problem was rather one of splitting the *nucleus* of the atom. 'Atomic' bombs should have been called 'nuclear' bombs: for the shattering energy they released came from the rupturing of nuclei (one nucleus, two or more nuclei).

Uranium

What makes uranium so dramatically different from other substances? To appreciate its unique characteristics we must first consider some basic 'nuclear physics': that is, what nuclei consist of and how they behave. An atom is made of electrons around a nucleus; a nucleus in turn is made up of 'protons' and 'neutrons'. A proton has a positive charge; a neutron has no electrical charge, and is 'neutral'. At first it seems difficult to understand how a nucleus stays together at all. The positive charges of the protons ought to push them violently apart. But within the compact volume of the nucleus a new kind of force comes into effect: an immensely powerful short-range attractive force acting equally between protons and neutrons – which, from this point of view, are all 'nucleons'. The short-range nuclear force holds them together, against the repulsive effect of the protons' positive charges. In this way the neutrons act as 'nuclear cement'.

However, in a nucleus which contains 92 protons – that is, a nucleus of uranium – the repulsive force among the protons is on the verge of overcoming the nuclear force. If there are as many as 146 neutrons also present, the nucleus can remain intact – barely. This form of uranium containing in all 238 nucleons is called uranium 238 or $^{238}_{92}U$. For reasons that need not concern us here, involving the grouping and compatibility of nucleons, the next most probable arrangement is a uranium nucleus containing three fewer neutrons: uranium 235, $^{235}_{92}U$. Atoms with these lighter nuclei make up about 0.7 per cent of naturally-occurring uranium. (If nuclei have the same number of protons, they are nuclei of the same 'element': thus, every nucleus with 92 protons is in an atom of uranium. Atoms whose nuclei have the same number of protons but different numbers of neutrons are called 'isotopes' of the element: for instance, uranium 238 and 235.) The uranium 235 nucleus has a property unique among all the more than 200 types of nuclei found in nature before 1942. The uranium 235 nucleus is under such internal stress that, as its nucleons churn about, sooner or later they abruptly rupture the nucleus completely.

Radioactivity produces radiation

The rupture of a uranium 235 nucleus usually results in about two-fifths of the original nucleus flying off in one direction and about three-fifths in the opposite direction, with perhaps two or three odd neutrons also shooting out. The flying fragments burst out with so much energy that a subsequent tally of masses reveals a shortage: some of the mass of the original nucleus has been converted into energy. This is the source of the enormous energies released in such nuclear events.

A possible sub-division might see one chunk including 38 protons and 52 neutrons, another including 54 protons and 88 neutrons, and three odd neutrons: making up, of course, 235 nucleons in all. The chunk containing 38 protons is strontium; since it contains in all 90 nucleons it is the notorious strontium 90. The chunk containing 54 protons is the inert gas xenon; since it contains in all 142 nucleons it is xenon 142.

Such a complete rupture of a nucleus is called a 'fission', by analogy with the biological term for the division of a growing cell. When it happens to a nucleus it is called 'nuclear fission'. It is the most violent kind of breakdown that can affect a nucleus. But there are others. A nucleus of uranium 238, for instance, while not being so near to rupture as its lighter relative, is still under severe stress: so much so that sooner or later it is likely to squirt out a lump containing two protons and two neutrons. Since this makes a larger proportional reduction in the proton contingent than in the neutron contingent, the remaining nucleus, now containing only 90 protons and 144 neutrons, is slightly less stressed. (It is the metal thorium 234.) The lump or 'particle' squirted out is identical in every respect to an ordinary helium nucleus; but since it emerges with considerable velocity, and ploughs a furrow through whatever slows it down, it is given a special name: an 'alpha particle'. Only nuclei with at least 83 protons undergo a breakdown this violent; they are called 'alpha emitters'.

The balance between protons and neutrons in thorium 234, while more satisfactory, is far from ideal. In effect, by emitting an alpha particle, the nucleus has over-adjusted. This leads to a yet more delicate form of breakdown. Out of the nucleus containing 90 protons and 144 neutrons there suddenly squirts an electron. It is identical in every respect to the electrons outside the nucleus; but since it emerges with considerable velocity it too is given a special name: a 'beta particle'. The nucleus which remains now contains one more positive charge than it had. But since an electron is very much less massive than a nucleon, there are the same number of nucleons as before – 234. A neutron has apparently turned into a proton. The nucleus now contains 91 protons and 143 neutrons: protactinium 234, another beta emitter, which becomes uranium 234, another alpha emitter. So, leap-frogging down, the nucleus alters itself until it has only 82 protons and 124 neutrons, and is at last stable: a nucleus of lead 206.

On the way, the nucleus regularly finds itself, after emitting an alpha or beta particle, still unduly agitated or 'excited'. To settle itself down it will then give off a burst of energy in a form closely akin to ordinary light, but much more energetic, and invisible. This burst of energy is called a 'gamma ray'. It is identical in every respect to the well-known 'X-ray', except that an X-ray comes from the electron-layers outside the nucleus, whereas a gamma ray comes from inside the nucleus.

Consider also the nucleus of strontium 90, one of the two large fragments formed in the example of fission mentioned earlier. The strontium 90 nucleus, a so-called 'fission product', has a disproportionately large number of neutrons for its protons, coming as it does from a much heavier nucleus which requires more 'cement'. Accordingly, the strontium 90 nucleus is also a 'beta emitter'. Sooner or later it will squirt out a high-velocity electron – a beta particle – and one of its neutrons will be replaced by a proton. Some fission products also emit a gamma ray or two.

There are thus four ways in which a nucleus can alter itself: fission, alpha emission, beta emission or gamma emission. Since, from a lump of such nuclei, the emissions from this activity are shooting out radially in all directions, the lump is 'radioactive'; and the emissions – neutrons, alpha and beta particles and gamma rays – are called 'radiation'. A collection of nuclei which shoot out 37000 million of such emissions in one second is said to have one 'curie' of radioactivity. (This is the radioactivity of one gram of radium, which was discovered by Marie Curie.)

In a radioactive substance it is impossible to tell whether a particular nucleus is on the point of radioactive breakdown, or 'decay'. Nonetheless, in a sufficiently large sample of any particular radioactive nuclear species, or 'radioisotope', a certain fraction of the nuclei will always decay in a quite regular length of time. For instance, if you start with 1000 nuclei of strontium 90, 28 years later 500 will have decayed and you will have 500 left. After a further 28 years, of this 500, 250 will have decayed, and you will have 250 left. And so on; however much you start with, 28 years later half will have decayed and only half will be left. Obviously the corresponding radioactivity will also have fallen by one-half. For strontium 90 the period of 28 years is called its 'half-life'. Each radioisotope has a half-life for each form of radioactivity it exhibits. Half-lives for different radioisotopes range from fractions of millionths of a second to millions of years.

The effects of radiation

Unless radioactive decay takes place in a vacuum, the radiation emitted must pass through the surrounding substance. The consequences depend on the substance, on the type of radiation, and on its energy. An alpha particle, made up of four nucleons with two positive charges, interacts vigorously with surrounding atoms, tearing off electrons and knocking nuclei out of place. In doing so, the alpha particle quickly gives up its own energy, travelling only a short distance but doing enormous damage along its path. A beta particle, much less massive and with only one negative charge, disturbs and dislodges neighbouring electrons, but loses its energy less swiftly and travels somewhat farther than an alpha particle. A gamma ray, with no electrical charge, loses its energy much more gradually, and can travel a long distance, causing a relatively small amount of disturbance at any particular point on its path. A neutron, also without electrical charge, is likewise free to travel a long distance, and is slowed down mainly by direct collision with nuclei.

Dislodging an electron from an atom makes the atom an ion: so emissions from nuclei are 'ionising radiation'. The effects of ionising radiation become particularly important if the radiation is passing through living matter. The delicate molecular arrangements of living matter can be easily upset by radiation; the effects depend roughly on how much energy is released into the living matter to disarrange it. There are several units used to measure radiation effects on living matter. The most common are the 'radiation absorbed dose', or 'rad', and the 'roentgen equivalent man', or 'rem'; the latter allows for the greater severity of alpha or neutron damage for equivalent energy-delivery. For beta and gamma radiation one rad is about the same as one rem; for neutrons and alpha particles one rad may be up to 20 rem, depending on the energy of the particles.

As we shall see, the question of the biological effects of radiation is surrounded by controversy. But it is known that a dose of perhaps 400 rem of radiation over the whole body will kill half the adult human beings exposed to it; and very much smaller doses will produce cell damage that may lead to leukaemia and other kinds of cancer. Furthermore, radiation damage to the complex molecules in the reproductive cells which contain the hereditary information may produce mutant offspring. Even a single gamma ray can disrupt a gene; it may produce unforeseeable effects if this particular gene should be in a reproductive cell which subsequently helps to form a child.

For more information about radiation biology see the Bibliography; suffice it to say here that the danger of radiation to living matter seems to increase in direct proportion to the amount of radiation exposure, beginning from the very lowest doses. There does not appear to be a dose below which damage does not occur. We are already subjected to continual radiation from the natural radioactive substances in our surroundings, and from cosmic rays. Any human activity which tends to add further sources of radiation to our surroundings must be considered as potentially harmful. Just how harmful – and in return for what benefits – is still under debate; this book is intended to make one aspect of the debate more intelligible, whatever your criteria.

Induced fission

Suppose, now, that in a collection of uranium 235 nuclei one nucleus undergoes a so-called spontaneous fission. As well as the two fission products, there shoot outwards perhaps two or three high-energy neutrons. (The chances are better than 99 to 1 that these neutrons will emerge virtually at the instant of fission: 'prompt' neutrons. But there is a slight chance that a neutron will not emerge until some seconds later: a 'delayed' neutron. As we shall see, such delayed neutrons are of considerable importance.) There are three possibilities open to the high-energy neutrons from fission. A neutron may reach the surface of the material and escape. It may strike another nucleus and be absorbed without causing any immediate breakdown. Or – most importantly – it may strike another nucleus and, in turn, cause this nucleus to rupture. The chances of a neutron causing such an 'induced fission' depend on its energy, and on the nucleus it strikes. A fast neutron, fresh from an earlier fission, can cause a nucleus of uranium 238 to rupture; but such a fast neutron is not the most effective means of rupturing a uranium 235 nucleus. If a neutron ricochets among other nuclei, bouncing off each and giving up its energy bit by bit, it soon slows down until it is just jostling with the shared heat-energy of the rest of the material. It is then a 'thermal neutron'. Only a fast neutron can rupture uranium 238. But a thermal neutron is much more likely to rupture uranium 235.

The chain reaction

If, in a lump of uranium 235, one nucleus undergoes spontaneous fission, the neutrons it releases may strike other nuclei, causing induced fission and releasing more neutrons. If there are enough uranium 235 nuclei nearby, the spreading disruption will multiply with astonishing rapidity: more and more neutrons; more and more ruptured nuclei, their fragments flying; more and more energy: a 'chain reaction'. If there is enough uranium 235, packed closely together for long enough, and if the chain reaction is out of control, the result will be a nuclear explosion: an 'atomic bomb'.

Needless to say, as soon as such an arrangement became possible it was tried out – on 16 July 1945, at the top of a tall tower in the desert near Alamogordo, New Mexico. Within three weeks an atomic bomb made of uranium 235 devastated Hiroshima. But, seemingly, one 'doomsday weapon' was not enough, and uranium 235 was not the only nucleus that could be used. A neutron can penetrate a nucleus of uranium 238 without rupturing it. If this happens, the resulting neutron-heavy nucleus soon emits a beta particle, and then another, to become plutonium 239. Like uranium 235 – and unlike all but a few other isotopes, all very rare – plutonium 239 is 'fissile': that is, it will rupture spontaneously into two lighter fission products plus odd neutrons. Because of the odd neutrons, plutonium

239 will also support a chain reaction. On 9 August 1945 such a chain reaction obliterated Nagasaki.

The nuclear reactor

If chain reactions in uranium 235 and plutonium 239 could be used only in weapons, the situation would already be sufficiently complicated. But more than two years before enough fissile material of either kind had been accumulated to make a weapon, it was found possible to control a chain reaction: to have it maintain itself without running away. Indeed it was by this means that the plutonium was produced for the Nagasaki bomb. The arrangement used to create and control a sustained nuclear chain reaction is called a 'nuclear reactor'.

The difference between an uncontrolled and a controlled chain reaction is profound. An arrangement of fissile nuclei which is to undergo an uncontrolled chain reaction – a nuclear explosion – must be sudden and final. An arrangement of fissile nuclei which is to sustain a continuing controlled chain reaction must be much more carefully organised. Curiously enough it takes many more nuclei – that is, much more material – to build a reactor than it takes to set off an explosion. The reason for this is the role played by the all-important neutrons.

If a chain reaction is to be self-sustaining it must keep itself supplied with neutrons. Consider the following sequence. A neutron plunges into a nucleus of uranium 235. The nucleus ruptures; as well as two fission-product nuclei, it also shoots out three neutrons. One of these three goes out through the surface of the lump of uranium and is lost. Another is absorbed by a nucleus of uranium 238, which begins its two-stage change into plutonium 239 but does not rupture. This leaves one neutron. If this third neutron now plunges into another nucleus and ruptures it, the process can continue; otherwise the chain reaction is snuffed out.

At any instant, inside the lump of uranium, there must be enough neutrons of the proper energy to propagate the chain. In effect, for a sustained chain reaction each neutron which is lost by causing a fission must be replaced by exactly one neutron which does likewise. When this condition is achieved the system is said to be 'critical', and the situation is called 'criticality'. (Note that 'criticality' does not necessarily imply danger.) If on average each neutron which is lost is replaced by more than one which also causes fission, the reaction will 'run away'; if on average each neutron so lost is replaced by fewer than one which causes fission, the reaction will stop. This is why a piece of uranium below a certain minimum size cannot, under any circumstances, sustain a chain reaction itself: there is too much surface for neutrons to leak out.

Moderators

The basic requirements for a continuing, controlled chain reaction are therefore (i) a collection of fissile nuclei appropriately distributed in space; and (ii) a self-replenishing supply of neutrons of just sufficient numbers and energy to keep the chain reaction going. In natural uranium, only 0.7 per cent of the nuclei are fissile uranium 235. These fissile nuclei are not sufficiently close together to keep up a chain reaction; too many neutrons are absorbed by the heavier uranium 238 nuclei. To improve the prospects for a sustained chain reaction it is necessary either to (i) increase the proportion of uranium 235 relative to uranium 238; or (ii) slow down the neutrons to thermal energies, at which they are much more readily absorbed by uranium 235; or (iii) do both.

As we shall see in Chapter 4, increasing the proportion of fissile uranium 235 – so-called 'enrichment' of the uranium – is a complex and expensive process. But even a small increase, say from 0.7 per cent to 2 or 3 per cent, makes a marked difference: provided that a means is also provided to slow down the neutrons from fission. Such a means is called a 'moderator'. A fast neutron, fresh from a ruptured nucleus, will usually bounce off a heavy nucleus like a bullet off a rock, losing hardly any energy. But if a fast neutron strikes a light nucleus, the neutron will give up a fraction of its energy, and after a few such collisions will have slowed to thermal energy. Best of all as moderators are the lightest nuclei: those of hydrogen. An ordinary hydrogen nucleus is, in fact, just a proton – almost the same mass as a neutron. Ideally, in a head-on collision the fast neutron would stop dead and the proton would shoot away with the energy. There is, however, a complication with ordinary hydrogen nuclei: a neutron which hits the proton is likely to stick to it instead of bouncing off, thus removing the neutron from the chain reaction.

Nonetheless, ordinary water, containing two hydrogen atoms per molecule, is a satisfactory moderator. But a rarer form of hydrogen nucleus is the best possible moderator. This rarer form is made up of one proton and one neutron: exactly the form created if an ordinary hydrogen nucleus absorbs a neutron. The proton-plus-neutron form of hydrogen is called 'heavy hydrogen' or 'deuterium'; its nucleus is called a 'deuteron'. If two atoms of heavy hydrogen combine with an atom of oxygen, the result is a molecule of 'heavy water', or deuterium oxide, which is much the best moderator of fast neutrons.

In addition to ordinary water and heavy water one other substance is widely used as a moderator: carbon, in the form of graphite. A carbon nucleus – six protons and six neutrons – is much more massive than either form of hydrogen nucleus, and is therefore not such a good moderator. But a carbon nucleus is not so likely to absorb neutrons as ordinary hydrogen, nor is graphite as expensive as heavy water. So carbon is also a popular moderator.

Reactor design

To set up a nuclear reactor, you proceed as follows. You take a good many pieces of material containing uranium 235 – usually uranium metal or oxide, natural or enriched – the 'fuel'. For a large reactor you will probably need a good many tons. You seal each piece into a casing called 'cladding', to confine the fission products that will be produced. You position the pieces, called 'fuel elements', in space, supporting them as necessary; remember that they are very heavy indeed. You intersperse the fuel elements with moderator, arranging the geometry of fuel and moderator so that there will be just enough slow thermal neutrons continually available to sustain the chain reaction – provided that you count both prompt neutrons and delayed neutrons. As long as the delayed neutrons are necessary to sustain the reaction you can control its speed. If you want to slow down the reaction, you insert some material which is a good absorber of neutrons, such as cadmium or boron, and soak up enough neutrons to reduce the reaction-rate as desired. Conversely, if you want to speed up the reaction, you withdraw your neutron absorbers to allow more neutrons to participate in the chain reaction until its speed has increased as desired. The delayed neutrons allow you to adjust the reaction-rate gradually instead of abruptly. The neutron-absorbers may, for instance, be rods, called 'control rods', inserted into the reaction region, which is called the 'core' of the reactor. To economise on neutrons you can surround this core with a reflector of some material which will bounce errant neutrons back into the reaction region. The best reflecting materials are the moderator materials aforementioned; in effect you extend the volume of moderator beyond the region where you have the fuel elements.

Once you have pulled out the control rods to let your reactor 'go critical', you must take precautions against the outpouring radiation from the core. Neither alpha nor beta particles will get beyond the fuel cladding (unless it leaks), but gamma rays and neutrons can travel through several feet of concrete and still be dangerous to living matter. Therefore, you surround your reactor with enough concrete or other protective 'shielding' to cut down the radiation outside to as low a level as you think advisable.

If you have to stop the chain reaction quickly, in the event of a malfunction, for instance, you insert the control rods as rapidly as possible. This is called 'scramming' the reactor. Even if you do not shut down the reactor, its reaction-rate will gradually dwindle of its own accord; but not, as you might suppose, because the fissile nuclei are being used up. What happens is that some of the fission products created within the fuel elements are themselves very good absorbers of neutrons – without of course being fissile. The worst offender is the radioisotope xenon 135. By the time two or three per cent of the fissile nuclei in the fuel have undergone fission, there is so much neutron-hungry xenon present that the 'reactivity' of the reactor is seriously affected: the phenomenon is called 'xenon poisoning'. By this time it is necessary to take out the fuel elements, or at least some of them, and replace them with fresh ones; the process is called 'recharging' or 'refuelling'. Since by this time the reactor core is full of radioactive fission products, extreme precautions must be taken.

If you set up a reactor on a sufficiently large scale, and let the chain reaction run fast enough, the energy released by the rupturing uranium 235 nuclei will make the whole assembly hot – potentially very hot indeed. Complete fission of all the nuclei in a kilogram of uranium 235 would release a total energy of about one million kilowatt-days – that is, as much heat as would be given off by one million one-bar electric fires operating for one 24-hour day. That is a lot of heat. Accordingly, the fuel in a reactor must be arranged to keep the neutrons and the fissile nuclei less concentrated, so that the heat is given off gradually enough to keep temperatures manageable. The amount of heat given off per unit volume in a reactor core is called the 'power density'. It may be anything up to several hundred kilowatts per litre; if such an outpouring of energy is not to melt – and indeed boil – the whole aggregation of material, it must be efficiently cooled. (The build-up of fission products in the fuel is directly related to the amount of heat generated per mass of fuel. It is called the 'burn-up' of the fuel, usually expressed in kilowatt-days per metric ton or, more often [since 1000 kw is one 'megawatt'], in megawatt-days per metric ton – MWd/T.)

The reactor can be cooled by pumping a heat-absorbing fluid through the core, past the hot fuel elements. The fluid can be a gas such as air, carbon dioxide, or helium; or a liquid such as water or molten metal. The choice of coolant depends on how fast heat has to be removed; on how many neutrons the coolant will absorb; on how readily it will react chemically with fuel cladding and moderator; on how expensive it is; on how easy it is to pump – and so on. The coolant carries the heat out of the reactor core. If the purpose of running the reactor is mainly to generate neutrons with which to turn uranium 238 into plutonium 239 for nuclear weapons – as was the case with the first large reactors – the heat is just a nuisance to be got rid of. But with appropriate arrangements this heat, like the heat from burning coal or oil, can be made to generate steam to run turbines or other forms of electrical generators. Such an arrangement – a nuclear reactor providing heat to run an electrical generating plant – is called a nuclear power station, or (in the US) a nuclear power plant.

In the following chapters we shall look much more closely at the structure and operation of the several main types of nuclear reactors. Each type receives energy from the fission of nuclei; but each uses a different arrangement of these nuclei: different designs of fuel, of moderator, of cooling system, of controls, etc. The differences have many important implications, as we shall see.

REACTOR TYPES

Experimental and research reactors

The first nuclear reactor was constructed amid great wartime secrecy, in a disused squash court under Stagg Field football stadium at the University of Chicago. It was built by piling up 57 layers of graphite bricks, some of which enclosed balls of uranium or uranium oxide, into a roughly spherical configuration inside a wooden supporting structure. When finished, the pile was some 20 feet high. On 2 December 1942 Enrico Fermi called out instructions and a young physicist named George Weil slowly pulled out the last cadmium-lined control rod. Shortly after 2.30 p.m., the neutron-detecting instruments recorded a steadily rising increase in neutron density in the pile. The pile had 'gone critical'; the first self-sustaining chain reaction was taking place.

Even before this first true nuclear reactor was built, 30 'piles' of less than the necessary size were built and tested. Such assemblies, which cannot generate their own neutron supply without supplementary neutrons, are called 'sub-critical assemblies'. Since the early 1940s countless hundreds of sub-critical assemblies have been built and dismantled in many countries; and many true reactors have been built for experimental or research purposes. In the most popular design the core is at the bottom of a deep tank of water. The water, while acting simultaneously as moderator, reflector, coolant and shielding, nonetheless allows a direct view of the core while the reactor is critical. In no other reactor design is this possible.

Experimental and research reactors have a number of uses. To bombard a sample of material with neutrons, it is inserted through an appropriate channel into the core, and left there until it has received the desired bombardment. The intention may be simply to study the effect of neutron bombardment on the material – perhaps a material to be used in building a reactor. Or the intention may be to convert some of the sample's stable nuclei by absorption of neutrons into radio-isotopes, for medical, industrial, agricultural or research purposes. Some research reactors are designed to further the study of 'reactor physics' itself: neutron densities, temperatures, the production of plutonium 239 from uranium 238, the build-up of fission products, the effect of these fission products on the effectiveness of the reactor (the reactivity), and so on.

Obviously, research reactors are also important for the training of qualified scientists and technicians in the often extremely subtle and intricate – and potentially dangerous – characteristics of reactor design and operation. In this context it is fair to say that the great majority of reactors, of whatever size, that have thus far been built have been experimental: in that every new reactor modification and development has had to be designed and engineered on the basis of previous experience, which in this field is very often inadequate or irrelevant or both. The US Atomic Energy Commission went so far as to list *all* the reactors it licensed, for whatever purpose, as 'experimental' until 1971.

Plutonium production reactors

All uranium reactors produce plutonium, by neutron bombardment of uranium 238. The first large-scale reactors were built expressly for this purpose: to produce plutonium for nuclear weapons. A pilot model was built in 1943 at Oak Ridge, Tennessee: a cube of graphite with horizontal channels containing cylindrical slugs of natural uranium clad in aluminium, cooled by air blown through the channels. But the full-scale reactors, eventually numbering nine, each as big as a five-storey building, were built along the bank of the Columbia River, near the town of Richland, Washington, on a 570-square-mile reserve named after the small town of Hanford. The Hanford reactors also used natural uranium clad in aluminium, stacked in horizontal channels in a graphite core. The heat they generated required high-efficiency cooling, which was achieved by pumping water from the river directly through the pile and back into the river. After the war production reactors were built on the Cumberland coast of Britain, on the site of a disused ordnance factory which was renamed Windscale. These reactors once again used natural uranium clad in aluminium, stacked in horizontal channels in a graphite core. Problems in finding a suitable supply of water for cooling led to the choice of air as coolant, blown through the pile at high velocity and discharged through tall stacks. Eight of the Hanford reactors and both Windscale reactors have since been shut down; in the case of the Windscale Number One reactor the shutdown followed a serious accident (see Chapter 3).

If the purpose of a reactor is to produce fissile plutonium 239, the rate of plutonium production can be 'optimised' by choice of core geometry, at the cost of other performance characteristics. The reactor fuel must be changed at relatively short intervals – less than two years on average. Otherwise more and more of the plutonium 239 will be converted, by absorption of an additional neutron, into plutonium 240, which is not fissile and is virtually impossible to separate from plutonium. 'Reprocessing' (see Chapter 4) is an expensive and complex operation; doing so more often than is strictly necessary to maintain the reactivity of the reactor can only be justified within the remarkable elasticity of military budgeting.

Gas-cooled power reactors

Magnox Reactors

British experience with air-cooling of the Windscale production reactors led to the choice of gas-cooling for their 'power reactors' – that is, reactors supplying heat to run electrical generators. The first British power reactors were 'power' reactors only secondarily; they were optimised for production of weapons-plutonium to augment the Windscale output. Four reactors were built at Calder Hall, adjoining Windscale, and four more at Chapelcross in Scotland. They formed the cornerstone of Britain's nuclear power programme.

Instead of once-through air-cooling, each of these eight reactors has a closed-circuit cooling system. The moderator is once again graphite, but the fuel channels are vertical. The entire core is enclosed within a massive steel pressure vessel some 21 m high and 11 m in diameter, and supported on an open framework called a 'diagrid'. The fuel is once again natural uranium, but the cladding is a new alloy of magnesium, much less neutron-absorbent, called 'Magnox' – from which the entire generation of reactors using such fuel acquired their name. In each fuel channel is a vertical stack of six 1 m-long cylindrical fuel elements, sheathed in Magnox, herringbone or spiral-grooved, to improve contact with the coolant gas. The whole vessel is enclosed in a biological shield of concrete more than 2 m thick. Access to the channels for refuelling and servicing is from above, through holes in the horizontal roof of thick concrete shielding called the 'pile cap'. Carbon dioxide gas is blown into the pressure vessel from below. The gas passes up each fuel channel between fuel element and moderator; by the time it has reached the top of the channel its temperature is some 345°C. This hot gas then passes out of the pressure vessel via four huge ducts, out through the concrete shielding of the reactor, and down through the pipework of four tall 'heat exchangers' – a fancy word for boilers. In the heat exchangers the hot gas heats water to make steam to drive a bank of turbo-generators. Then the gas, now once again cooler, is ducted through four 'gas circulators, or pumps, back through the concrete shielding and back into the bottom of the pressure vessel to repeat its journey up through the reactor core. Each of the four sets comprising one heat exchanger, one gas circulator, and one-quarter of the reactor core is called a 'loop': this is a 'four-loop' design.

Each sector of the core also has channels for several types of control rods: 'black' high-boron steel rods to 'scram' the reactor and give swift shut-down, 'grey' rods which act more slowly, and 'bulk' rods which are gradually withdrawn to compensate for burn-up. The rods enter from above. The scram rods are held out on electromagnetic grapples, designed so that any reactor fault will shut off the magnets and let the rods fall into the core to halt the fission reaction.

The carbon dioxide coolant is maintained at a pressure of some 8 atmospheres. Accordingly, special provisions have to be made for changing fuel elements and other maintenance inside the pressure vessel. On the pile cap, the working area above the reactor, are mobile 'charging' or 'refuelling machines', massive and complex assemblies. To change the fuel in the reactor, the 'discharge machine' is positioned over an access port, clamped on to the surface of the pile cap and pressurised to match the coolant pressure inside the reactor. The shielding plug is removed, grapples extended down into the core, and the irradiated fuel elements lifted and stored inside the thick walls of the discharge machine – all by remote control, because of the radiation hazards. The shielding plug is replaced, the discharge machine depressurised and moved, and the charge machine, loaded with fresh fuel, moved into position.

The whole cycle of clamping, pressurisation, unplugging – this time to lower new elements into place – replugging, depressurising and unclamping is repeated – once again by remote control. The 'irradiated' fuel, intensely radioactive with fission products, is then moved, inside the discharge machine, to be dropped into a 'cooling pond': a deep tank of water, which serves to shield and cool the fuel while the more short-lived fission products within it decay to a less dangerous level of activity. After some weeks the used irradiated fuel will then be transported to Windscale for 'reprocessing' – see Chapter 4.

The Calder Hall design served as the basis for the first generation of British commercial nuclear power stations. In all, eight Magnox stations, each with two identical reactors, were built for the Central Electricity Generating Board, and one for the South of Scotland Electricity Board. Design details varied from station to station (*Reactors UK* – see Bibliography). Reactor sizes increased, until for the last two CEGB stations a major modification took place. Instead of a steel pressure vessel, each of these twin-reactor stations was built using pre-stressed concrete pressure vessels, which could be built much larger than steel could be reliably welded; so that it became possible to enclose not only the reactor core but also the boilers and pumps. In this design, the pre-stressed concrete serves both as pressure vessel and as biological shield; the gas ducts are eliminated, removing one of the major sources of possible hazard that might arise from an accident (see Chapter 3). The fuel charges of the Magnox stations vary from some 110 metric tons of natural uranium for each

Calder Hall reactor up to 595 metric tons for each reactor at Wylfa, the newest and largest Magnox station. The power density of the Magnox stations averages about 0.9 kilowatts per litre. The 'specific power' – power per unit mass of fuel, which is closely related to the power density – varies from 2.40 kilowatts per kilogram at Calder Hall to 3.16 kilowatts per kilogram at Wylfa. The heat output per reactor varies from 268 megawatts – that is, 268,000 kilowatts – at Calder Hall to 1875 megawatts at Wylfa. This is converted in the turbo-generators to, respectively, some 50 megawatts and 590 megawatts of electricity. Note the increased conversion efficiency. A Calder Hall generator converts less than 20 per cent of the reactor-heat to electricity; a Wylfa generator converts more than 30 per cent.

The power output from a reactor can be expressed directly as heat – 'megawatts thermal' or MWt; or as ultimate electrical output – 'megawatts electric' or MWe; which is usually only one-third to one-quarter as large, because of the inefficiency of the conversion process from heat into electricity.

Among other major nuclear power programmes, that of France is the only other one to opt initially for gas-cooled reactors. The small French reactors at Marcoule and Avoine started up in 1958. The second unit at Avoine – Chinon-2, a 200 MWe reactor – went critical in 1964, and France has since built, in all, seven gas-cooled reactors with graphite moderators and a 70 MWe gas-cooled reactor with heavy-water moderator. But French interest has recently swung markedly away from gas-cooled to light-water designs in partnership with American reactor-builders. (See later in this chapter.)

Advanced Gas-Cooled Reactors

The low efficiency of thermal to electric conversion follows mainly from the comparatively low temperature of the coolant which emerges from the reactor core and generates steam in the boilers. If the steam could be made hotter, the proportion of useful energy could be considerably increased. But in the Magnox design the coolant temperature must not be allowed to go too high, because of the possibility of melting the Magnox fuel cladding; the uranium metal, too, may become distorted. So the second generation of British power reactors – the advanced gas-cooled reactors or AGRs – use a different design of fuel: uranium dioxide, clad in stainless steel. The dioxide has a much higher melting point, as does the stainless steel. But the stainless steel is much more prone to absorb neutrons than is Magnox. To offset this, the uranium must be slightly 'enriched': the proportion of fissile uranium 235 must be somewhat increased, from 0.7 per cent to about 2 per cent (see Chapter 4). To improve heat-transfer from fuel to coolant – again carbon dioxide – the stainless steel tubes containing the uranium dioxide are only about 1 cm in diameter; so-called 'fuel pins'. These pins are assembled in a cluster held apart by spacing grids and encased in a hollow graphite cylinder. The coolant passes through the hollow graphite.

The AGR design once again uses a pre-stressed concrete pressure vessel enclosing core, boilers and circulators. Access is from above, through the pile cap. Control rods enter the core from above. Refuelling is done with a single machine which both charges and discharges. The fuel elements, stacked eight to a vertical channel, are coupled together, removed and inserted as a single long string; accordingly the refuelling machine is the height of a four-storey building. The specific power of the first commercial AGR, at the Dungeness B Station, is to be 9.5 kilowatts per kilogram of uranium, some three times that of the largest Magnox station. The power density will be about 2.2 kilowatts per litre; the coolant temperature leaving the core, 675°C. But here it must be said that these are design ratings; the only AGR which has yet operated is the much smaller prototype at Windscale.

High temperature gas-cooled reactors

To improve thermal to electric conversion efficiency still further an even higher coolant temperature is desirable. This, however, involves a totally new engineering approach, in which the entire high-temperature region of the reactor contains no metal whatever, but only ceramic materials. The only high-temperature gas-cooled reactor – HTGR – thus far in operation is the Dragon reactor at the UK Atomic Energy Authority reactor development site at Winfrith, in Dorset. The Dragon is a prototype HTGR, built and operated as an international cooperative project under the OECD.

Fuel for the Dragon reactor is made in particles like

small shot. In the centre of each particle is a lump of uranium dioxide, highly enriched – up to as much as 93 per cent uranium 235. This lump is coated with a layer of baked carbon, a layer of silicon carbide, and a second layer of baked carbon, so that a magnified cross-section of the coated particle looks like a broken gob-stopper. These particles are fused together into bracelet-shapes called 'compacts' and stacked into a graphite sleeve; the sleeve full of compacts constitutes the fuel element. The coolant, which in the HTGR is not carbon dioxide but helium, passes up through the hollow centre of the element. The moderator is once again graphite. The specific power is of the order of 1.4 MW/kg; recall that the specific power of Dungeness B is only 9.5 kw/kg.

The Dragon reactor has been used primarily as a test-reactor for different designs of fuel, and is not coupled to an electrical generator. But at Fort St. Vrain in Colorado another design of HTGR, built by Gulf General Atomic, is being used in a prototype generating station which is expected to be in operation this year. The Fort St. Vrain design again uses coated-particle fuel, but it is fused into a different geometry. Even before starting-up of their prototype Gulf have already sold at least six full-scale reactors to other buyers.

A distinctive feature of the HTGR is that it does not need control rods. The neutron behaviour in the core is such that if the coolant gas is circulated faster, the resulting drop in core temperature changes the neutron flux as to boost the power output and maintain the temperature: whereas if the coolant is circulated more slowly the resulting rise in core temperature changes the neutron flux as to reduce the power output and suppress the temperature rise. The HTGR is said to have a 'negative temperature coefficient of reactivity'. Other reactor designs sometimes exhibit this inherently stable characteristic – but not through their entire performance repertoire, as does the HTGR.

A radically different design of HTGR is being developed in Germany. In this design, called a 'pebble-bed reactor', the core is a huge bin, filled with ceramic fuel shaped into what looks like black billiard balls. The balls incorporate both the fissile material and the carbon moderator. Helium coolant is blown up through the bin-full of balls. As the reactor operates the balls are shaken slowly down the bin, emerging at the bottom when they have reached their irradiation limit; new balls are fed in at the top. But this design further complicates one of the more difficult problems of the more orthodox HTGRs, in that the ceramic fuel in this form may be even more difficult to reprocess than that of the Dragon or Fort St. Vrain designs: and a leaking fuel ball would be effectively irretrievable.

Light-water reactors

Pressurised-Water Reactors

Like the first British power reactors – built to produce weapons-plutonium – the first US power reactors also began under military auspices, albeit specifically as power plants. The US Navy realised early in the 1950s that a submarine powered by nuclear fuel would not need to resurface to replenish oxygen supply, since the 'burning' of such fuel – unlike that of oil – did not require oxygen. Spurred by this idea, and constrained by the space limitations in a submarine, US designers developed a reactor which used a core of relatively high-power density, with fuel elements immersed in a tank of ordinary water – 'light water', to distinguish it from heavy water – under sufficient pressure to keep it from boiling. The water, contained in a steel pressure vessel and ducted through heat exchangers to produce steam for turbo-generators, served as coolant, moderator, and reflector. Its relatively high neutron-absorption required that the fuel be enriched (see Chapter 4); and the high-power density required special provisions for possible malfunctions (see Chapter 3). But the essential simplicity of the system has been a factor in making the pressurised water reactor, or PWR, by the 1970s the world's most popular reactor design. Since it first came ashore at Shippingport, Pennsylvania, in 1957, in the first US commercial nuclear power plant, the PWR has proliferated until it now outnumbers all other designs. There are now 76 PWRs in operation or under construction in 13 countries, and many more planned.

PWR fuel is uranium dioxide, typically enriched to about 3 per cent uranium 235, clad in a zirconium alloy called zircaloy. The fuel pins are nearly 4 m long, slender (about 1 cm in diameter) and closely spaced, less than 2 mm between pins. They are held by spring fingers in grids containing clusters of perhaps 17 by 17 pins, supported above and below on plates fixed to the pressure vessel with apertures for coolant flow. The power density may be some 65 kilowatts per litre, compared to about 1 kilowatt per litre in a Magnox reactor. Control rods and their drive mechanisms are included within the reactor pressure vessel, in the space above the core. The commonest PWR is a four-loop design, with heat exchangers and coolant pumps outside a thick concrete biological shield. Refuelling is carried out 'off load' – that is, with the reactor shut down. It is allowed to cool, then the top portion of the pressure vessel is unbolted, and the space above it flooded with water to provide shielding and cooling. About one-third of the fuel charge is replaced at each annual refuelling.

Some PWRs now under construction are intended to deliver more than 3000 megawatts of heat. But the relatively low coolant temperature attainable using water under managable pressure (some 150 atmospheres) makes the PWR a comparatively inefficient design for electrical generation. Usually less than 30 per cent of the heat is converted to electricity, compared for instance to a modern fossil fuel plant which may convert upwards of 40 per cent.

Boiling-Water Reactors

US interest in water cooling stemmed from the Hanford reactors and was furthered by the submarine PWRs. It was known that water allowed to boil is more effective in removing heat, but boiling was thought likely to trigger instabilities in a reactor core. However, research led to a design which seemed manageable, and which is by far the simplest of all power reactors, bearing a close resemblance to an ordinary steam-boiler. Like the PWR, the boiling-water reactor, or BWR, consists essentially of a tank of ordinary water surrounding a bank of vertical 'flues'. But these flues, instead of carrying hot gas to boil the water, contain nuclear fuel. As in the PWR, the water serves as moderator, coolant and reflector – and in addition, when boiled, provides steam which is ducted directly to drive turbo-generators. Once through the turbines, the coolant water is condensed and pumped again into the 'boiler' – that is, the reactor pressure vessel.

BWR fuel, like PWR fuel, is uranium dioxide, enriched to some 2.5 per cent uranium 235, and clad in zircaloy. The fuel pins are not quite so slender as those of a PWR, and are grouped, typically, in a 7 by 7 array or 'box', open-ended to allow coolant to pass through, and supported above and below on plates fixed to the pressure vessel. Since the space above the BWR core must include steam-collecting apparatus, the BWR control rods are inserted from below. Each control rod has a cross-section like a plus-sign whose arms are half the breadth of a fuel box; the control rods enter at the corners between fuel boxes so that each fuel box is effectively enclosed by four right-angles of control rod surface. Like a PWR, a BWR is refuelled off load. The top of the pressure vessel is unbolted and the space flooded with water to provide shielding and cooling. Care must be taken to avoid accidental criticality as fuel boxes are lifted out; the flooding water may be 'poisoned' by adding a neutron-absorbing boron compound.

Since the steam fed to the turbines comes directly from the BWR core, it may carry with it some radioactivity, making turbine maintenance somewhat more difficult. But in practice, most radioactive material in the BWR coolant stays in the liquid water and does not get carried over by steam into the turbine. The BWR shares with the PWR the drawback of comparatively low coolant temperature and resulting inefficiency of conversion of heat to electricity. It also shares the problems associated with relatively high power density, although the power density of a BWR is likely to be only half that of a comparable PWR, perhaps 35 kilowatts per litre. The BWR is also more potentially susceptible than the PWR to the possibility of 'burn-out' or 'steam blanketing', which arises if a layer of steam is allowed to form next to the hot fuel-cladding. The low heat-conductivity of the steam would mean that the heat would no longer be so effectively removed from the fuel, and the fuel temperature might then rise suddenly and dangerously (see Chapter 3). Design and operation of all types of reactor must take into account the possibility of sudden surges, called 'transients'. This is particularly true for the high power density light-water reactors, in which a temperature transient might lead to 'burn-out'.

Heavy water reactors

CANDU Reactors

The Canadian role in wartime fission research had been particularly concerned with heavy water. But after the war, Canada decided against a nuclear weapons programme. Accordingly, with no facilities for enrichment (see Chapter 4) but with substantial supplies of uranium, Canada chose to concentrate on heavy-water natural-uranium reactors. After a long period devoted primarily to research, interest led to development of a power reactor: the CANDU reactor (from CANadian Deuterium Uranium).

The basis of the CANDU reactor is a horizontal cylindrical tank with hundreds of horizontal tubes running completely through it from one end to the other, very like an ordinary boiler. This tank, called a 'calandria', is filled with heavy water at atmospheric pressure, kept well below boiling point by its own cooling circuits. The barrel and ends of the calandria are of stainless steel; the tubes are of zirconium alloy. In these calandria tubes are zirconium alloy pressure tubes containing natural uranium dioxide fuel rods in short (50 cm) clusters. The pressure tubes are connected individually to inlet and outlet feeders. The coolant, also heavy water, passes through the zirconium pressure tubes. The space between a pressure tube and the calandria tube enclosing it is filled with carbon dioxide to insulate the calandria full of moderator from the hot coolant tubes. The heavy-water coolant at some 90 atmospheres pressure passes through heat exchangers and pumps and generates steam to drive turbines; the system is here exactly analogous to that in a PWR with pressurised light-water coolant.

The reactor power is varied by varying the level of moderator in the calandria: less moderator means fewer fissions. The reactor can be scrammed by dumping all the moderator through large valves into a tank below the reactor; the valves are normally held shut by helium pressure, and any reactor fault will de-energise them, allowing them to open.

Refuelling is done 'on load', by means of machines which couple on to either end of a pressure tube. One machine pushes new fuel in one end, which in turn pushes irradiated fuel out through the other end into the other machine.

Heavy water is very expensive (from £13 per kg) so losses must be kept to a minimum. Each pressure tube has two mechanical seals, and leakage is a major problem with coolant at high temperature and pressure.

Steam Generating Heavy Water Reactors

The British Steam Generating Heavy Water Reactor – SGHWR – of which only a prototype exists (at Winfrith in Dorset) combines the features of the CANDU reactor and BWR. In the SGHWR a calandria filled with heavy-water moderator surrounds vertical pressure tubes; in each pressure tube is a single 4-m element – a cluster of zircaloy-clad enriched uranium dioxide fuel pins through which ordinary light-water coolant is circulated. The light water is allowed to boil, as in the BWR; the steam is circulated directly through turbogenerators, before being condensed and pumped again from below into the forest of pressure tubes. As in the CANDU reactor the SGHWR power level is varied by varying the level of the moderator; the SGHWR can be scrammed by automatically pouring boron solution into interstitial tubes in the calandria. Compensation for fission product build-up is made by gradually reducing the amount of boron compound dissolved in the moderator.

Like other light-water designs the SGHWR has a comparatively high power density. SGHWR size is unusually flexible; the combination of highly efficient heavy-water moderator and pressure tube design makes it feasible to build an SGHWR smaller than is physically manageable for most other designs.

Fast breeder reactors

All the reactors so far described share a common feature. For reasons of heat extraction and neutron economy they rely primarily on the energy released by fission of uranium 235 by slow 'thermal' neutrons; they can be called, as a group, 'thermal' reactors. Even in a thermal reactor, it is true, a small proportion of the available neutrons are absorbed by uranium 238, and the resulting fissile plutonium 239 makes a small contribution to the total release of energy. But the amount of plutonium created is much less than the amount of uranium 235 used up; so such reactors can also be called 'burner' reactors. Material like uranium 238, which can be converted, by absorption of neutrons, into fissile material like plutonium 239, is called 'fertile'. (Thorium 232 is also fertile; it can be converted by neutron bombardment into fissile uranium 233. But as yet for reactor design this possibility remains theoretical.) The comparison between fissile nuclei consumed and fertile nuclei converted to fissile is called the 'conversion ratio'; for instance, if for every uranium 235 nuclei undergoing fission one uranium 238 nucleus were converted to plutonium 239, the conversion ratio would be 0.1. In a burner reactor the conversion ratio is always less than 1. But it is possible to design a reactor in which this ratio is greater than 1: a 'breeder' reactor, which produces more fissile nuclei than it consumes.

As it happens, the first reactor ever to power electric generating equipment was such a reactor, the Experimental Breeder Reactor-1, or EBR-1, which in 1951 at the AEC reactor development installation at Arco, Idaho, produced some 100 watts of electricity. The first true power reactors to use this principle, however, were the British Dounreay Fast Reactor – DFR – near Caithness on the north Scottish coast, and the Detroit Edison Enrico Fermi 1 reactor near Detroit, Michigan. The DFR has been in operation since 1959; the Detroit Edison reactor however has experienced endless trouble, including an accident which might have necessitated the evacuation of Detroit (see Chapter 3).

To achieve a conversion ratio of better than 1 – that is, to 'breed' fissile nuclei faster than they are consumed – requires a high neutron flux. This is most readily achieved in a system with a high concentration of fissile nuclei and no moderator. Although a fast neutron does not so readily cause a fission, when it *does* the result produces on average more free neutrons than fission by a slow neutron. Furthermore, if the neutrons are not slowed down, non-useful absorption of neutrons is also reduced. Provided the fuel is sufficiently enriched – to at least 20 to 25 per cent fissile material – the neutrons do not get a chance to slow down before they produce the desired effects: either by causing fission – of either uranium 235 or uranium 238, or plutonium 239 – or by converting uranium 238 into plutonium 239. Since the arrangement depends on the role played by fast neutrons, it is known as a 'fast breeder'.

The main engineering problem presented by a fast breeder is that of cooling. The core is much more compact than that of a thermal reactor, with power density of the order of 350 to 450 kilowatts per litre. Neither gas nor water is sufficiently efficient to remove so much heat from such a small volume. Instead, the favoured coolant is liquid – that is, molten – metal. In the DFR the coolant is a molten alloy of sodium and potassium; but in more recent designs like the Detroit Edison reactor and the British Prototype Fast Reactor, also at Dounreay, the choice has fallen on pure liquid sodium. It does not readily absorb neutrons, but those it *does* absorb make some of the sodium nuclei into radioactive sodium 24. As a result, the primary coolant loop becomes intensely radioactive; this necessitates a second sodium loop, with a heat exchanger inside the biological shielding, to pick up the heat from the radioactive primary loop and carry it outside the shielding to a second heat exchanger in which steam is generated. Sodium is opaque, making internal inspection of cooling systems very difficult; and sodium coolant must not, of course, be allowed to cool below its melting point (97.5°C) anywhere in the circuits, or it will solidify.

The fuel of the DFR is 75 per cent enriched uranium clad in niobium, in fuel pins only about 1 cm in diameter. The core containing this fuel is only 53 cm high, hexagonal, 52 cm across the flats. Around the core proper is a so-called 'breeder blanket', which contains fertile uranium 238 to catch the intense neutron flux emanating from the central core. Both core and breeder blanket are cooled by liquid sodium-potassium alloy, flowing at high speed. The coolant emerges from the core at 350°C, and passes through 24 heat exchangers inside the concrete biological shield, transferring heat to a secondary liquid-metal circuit which carries the heat out of the shield to a second set of heat exchangers producing steam to drive turbines. The maximum power output of the DFR is 60 megawatts thermal, 14 megawatts electric. The reaction-rate control and shutdown are provided by groups of movable fuel elements around the core. The whole reactor is enclosed within a steel containment sphere 41 m in diameter, of which the lower hemisphere is lined with concrete 1.5 m thick. The containment and the remote location were employed to give additional safety features during the development of fast breeder technology.

Much larger breeder installations are now complete or nearing completion in France (the Phénix reactor), in the USSR (the Shevchenko reactor) and at Dounreay (the Prototype Fast Reactor or PFR); the US has announced plans for a prototype installation on the Clinch River near Oak Ridge, Tennessee. The intention is that these larger fast breeders will be able to use not only uranium 235 but also plutonium 239 as fuel: perhaps 1 to 3 metric tons of plutonium per fuel charge. In this way, ideally, virtually the whole of the supply of uranium can ultimately serve as fuel, either directly as uranium 235 or indirectly via conversion to plutonium 239. As it happens, one of the currently limiting factors is the availability of plutonium for fuel charges. Optimistic estimates suggest that it may take a fast breeder reactor 10 years to double its inventory of plutonium by breeding; more realistic estimates make this 'doubling time' nearer 20 years. A single reactor needs not only enough plutonium for its operating fuel charge, but enough in addition to allow for cooling, transport, reprocessing and fabrication – in all, considerably more than the fuel charge itself: the 'pipeline inventory'. To get reactor plutonium, the various countries may have to cannibalise their arsenals of nuclear weapons. Some would call that the strongest vote in favour of the fast breeders.

3

REACTORS AND THE ENVIRONMENT

Reactor safety

Designers, builders and operators of nuclear reactors have laboured long and hard to dispel the public suspicion that a nuclear reactor could explode 'like an atom bomb'. It is worth saying here at the outset that under no circumstances could a thermal neutron reactor of whatever design cause a nuclear explosion. The fissile material is simply too dilute, and could not, by whatever mishap, become sufficiently concentrated (enrichment – see Chapter 4 – is complex and difficult for just this reason). However, like many other types of large industrial installation, a nuclear reactor could conceivably experience, as a result of drastic internal malfunction, a *non-nuclear* explosion. The consequences would be comparable in every way but one to those of such an accident in any industry: deaths, injury, property damage. What uniquely distinguishes a reactor from other installations is the radioactivity it contains, which in the event of an accident might be released.

After a large power reactor has been in operation for some months, the accumulated fission products in its fuel charge dwarf the amount of radioactivity released over Hiroshima. A 1957 study prepared at Brookhaven

National Laboratory by the US Atomic Energy Commission, called 'Theoretical Possibilities and Consequences of Major Accidents in Large Nuclear Power Plants', or WASH-740 as it is commonly known, predicted that the 'maximum credible accident' at their theoretical reactor would lead to 3400 deaths, 43 000 injuries and property damage of $7000 million. A study by the University of Michigan, using the same basis, forecast as many as 133 000 deaths. (The term 'credible accident' occurs frequently in reactor safety studies – the sparseness of operating experience to date does not give a statistical basis for evaluation beyond the assertion of the 'incredibility' of certain accidents which have not yet happened.) The AEC, for reasons known only to themselves, have since refused to publish an up-dated study of accident possibilities, although the reactors now operating and under construction will be at least an order of magnitude larger than that assumed in the 1957 study. (Other studies offer piecemeal suggestions. A 1970 report by the AEC consultants at the Battelle Memorial Institute, BMI-1910, indicates for one hypothetical accident the possibility that thyroid exposure from iodine 131 at a distance of 100 km from the reactor might be several hundred rads – probably fatal, certainly severely injurious.) Clearly, the radioactive contents of a reactor must not, under any circumstances, be allowed to escape.

Such an absolute requirement comes up against three obstacles: technological, economic, and human. Reactors of whatever kind are designed to confine the fission products no matter what credible accident happens. To begin with, the fuel itself, in which the fission products are generated, is hermetically sealed in cladding. Then the fuel elements are enclosed within the sealed volume of the reactor vessel and its cooling circuits. There will then be at least one further shell of so-called containment: say, the reactor building itself, which will be designed so that it can be completely sealed against the escape of gases, providing a third line of defence against release of fission products to the outside air.

It sounds encouraging. But the first line of defence – the fuel cladding – has manifested with disconcerting regularity a tendency to leak. When this happens, gaseous fission products get into the primary coolant (water, carbon dioxide, etc.). In addition, impurities in the coolant may, under the intense neutron bombardment, be transmuted into radioactive forms, as may corrosion products formed on the outside of the cladding, adding to the activity in the coolant (see later in this chapter). So the real 'first' line of defence is the reactor vessel itself.

If the coolant is pressurised, as it is in most thermal reactors to a greater or lesser extent, the integrity of the pressure system becomes of the utmost importance. The two most common designs are those which use welded steel (such as light-water reactors, the early Magnox reactors and the European HTGR) and those which use pre-stressed concrete (such as the later Magnox reactors and the AGRs). The heavy-water systems (like the SGHWR and the CANDU reactors) have the coolant passing not through a single large vessel but through a battery of hundreds of parallel pressure tubes.

A pre-stressed concrete pressure vessel is held in compression by thousands of steel cables, individually secured. The integrity of such a pressure vessel, more than 10 feet thick in every direction, against even the most violent phenomenon conceivable in its interior, seems unquestionable. The same is said, by both British and American reactor builders, about welded steel vessels, at least by implication: because for both industries the 'maximum credible accident' they consider is a double-ended break of the input coolant duct between pump and pressure vessel. But a steel pressure vessel is a huge barrel of welded steel, subjected to high pressures, high temperatures and intense neutron bombardment. Boilers in non-nuclear plants have been known to burst. Such boilers, while perhaps subjected to higher pressures or temperatures than a reactor pressure vessel, have not undergone intense neutron bombardment. The reactor vessel has. Neutron bombardment changes crystal structures; the distorting and destructive effects it produces on a smaller scale are well known but continue to reveal new aspects. Thus far, the long-term effects of intense neutron bombardment on large-scale structures are simply outside the realm of practical experience.

Be that as it may, the maximum credible accident considered by reactor builders is effectively an abrupt and total loss of pressurisation. It is assumed – and this is important – that the immediate consequence of this or any less serious accident is an automatic reactor scram: emergency insertion of control rods to shut down the fission reaction. (This is also known as a 'reactor trip'.) Emergency scram devices include, for instance, control rods suspended by electromagnets, so that the rods will drop into the core if the magnet current is shut off; sprays of boron solution or powder; and baskets of boron-steel balls electromagnetically suspended, used in graphite-core reactors in case core distortion should interfere with rod insertion. Nonetheless, there have been instances of scram failure – fortunately not in the context of a loss of pressurisation. When, as in such a case, a parallel set of identical safety devices all fail for the same reason (usually a design flaw), the result is called a common-mode failure. Records of reactor operating experience reveal many examples, in a wide variety of contexts – none serious, thus far.

Once a reactor is scrammed, and its fission reaction shut down, the majority of the internal heat generation is cut off – but not all. Some of the heat in the core comes not from fission but from the radioactivity of the fission products themselves; in gas-cooled reactors perhaps 6 per cent, in water-cooled reactors with higher power density perhaps 10 per cent or more. This 'decay heating' is unaffected by scramming the reactor, although it decreases as the fission products decay. Loss of pressurisation, even if followed within seconds by a reactor scram, means less efficient cooling, leading to a surge of temperature within the core. In a gas-cooled reactor the surge brings with it the possibility of melting or even ignition of Magnox cladding, coupled with serious distortion of the uranium metal; however, stainless steel cladding and uranium dioxide fuel in AGRs will have a margin of safety some hundreds of centigrade degrees above the probable temperature maximum. Gas-cooled reactors are provided with emergency blowers and emergency supplies of carbon dioxide; even the relatively slow circulation of coolant should be enough to keep the fuel below dangerous temperatures. The mass of graphite moderator itself tends to soak up excess heat. The main requirement is that air be excluded from the core; if air were to enter, it might lead to ignition not only of the fuel but also of the graphite. In 1957 the Windscale Number One plutonium reactor was destroyed by an internal fire caused when an unexpected surge of heat ignited fuel and graphite in the air coolant. Only filters, installed in the preceding months as a belated precaution on the coolant-discharge stacks, prevented a disastrous spread of radioactivity over the surrounding countryside. As it was, many thousands of gallons of milk contaminated with radioactive iodine 131 had to be poured into the sea.

Loss of pressurisation in a water-cooled reactor, with its higher power density, could have a much more alarming sequel. In the maximum credible accident – a break in a coolant inlet duct just outside the reactor vessel – the whole of the cooling water might be lost in a matter of seconds. Recall that the water is under very high pressure; such a loss-of-coolant accident or LOCA is often referred to as a 'blow-down'. In the first few seconds after blow-down the decay heating from the fuel will cause the core temperature to shoot up; if this surge of temperature is not arrested within 15 seconds the consequences may be very serious indeed. The zircaloy cladding may weaken or melt; the zircaloy may react with the water, releasing hydrogen which may cause a major explosion; the buckling fuel elements may totally block coolant flow; the resulting collapse of the core may lead to a major 'melt-down', in which the pool of intensely radioactive molten metal, still generating its own heat, plummets through both pressure vessel and containment, melting, burning and exploding its way downward under gravity, impossible to arrest. Such an eventuality has been sardonically termed, because of its direction of progress, the 'China syndrome'. The release of radioactivity caused by such an accident at a large modern light-water reactor could have consequences that would make even the Brookhaven figures look comforting.

To forestall such an outcome, light-water reactors are fitted with emergency core cooling systems, or ECCS. In a pressurised water reactor the ECCS is designed to flood the core with emergency cooling water from below; in a BWR the ECCS is designed to spray emergency cooling water from above. However, such systems have never been tested under accident conditions in a full-scale reactor; and, despite prolonged and expensive computer simulations and tests of models, there seems every possibility that the ECCS on both PWRs and BWRs may not work. The story is too long to tell here, and is far from over; for details see the Bibliography under ECCS.

The intense controversy over ECCS is only one of several themes relating to reactor safety, or the lack thereof, now coming increasingly into prominence. Another was revealed when fuel elements discharged from the Beznau Number One reactor, a PWR in Switzerland, and from the Robert Ginna reactor in

Rochester, New York, both proved to have undergone serious deformation during irradiation. The fuel pin cladding was crushed and crumpled, and upon examination was found to be partially empty inside. No reason for this development has thus far been reliably established; but fuel of a similar type is being used in several reactors now in operation in the US, and worries are being expressed that a wholly new phenomenon has been discovered, not necessarily confined to this type of fuel element. The AEC, belatedly informed of the discovery, reacted by issuing an edict that, in effect, nothing be done: that power levels of other reactors using the type of fuel in question be neither increased nor decreased. As edicts go it was comparatively easy to obey. Whether it meets the situation is more questionable.

Other accident possibilities are numerous, varying downward to the trivial. One worth mentioning is generally rated among the 'incredible'. PWRs and BWRs are refuelled off load; and the top of the pressure vessel, removable for refuelling, is otherwise held in place by a ring of huge bolts. It has been suggested that, in the event of internal malfunction or otherwise, the 'lid' of the pressure vessel might abruptly part company with its lower partner – in which case the massive lid, weighing many tons, would be blown skyward with ample force to penetrate whatever structures it encountered. Once again, the effect on the surrounding countryside would be unfortunate. But, returning to the realm of the 'credible', lesser accidents could involve, for instance: partial failure of coolant circulation, due to internal blockage or pump failure;

failure of control-rod drives; valve failure; electrical failures of many kinds; and – above all – simple human failure. As reactor operation becomes more and more routine, as personnel come to take it for granted, as earlier dedication gives way to everyday job-holding, the probability of operator error grows. There have already been spectacular instances; in 1970 the Dresden 2 BWR near Chicago spent several hours with its water-coolant falling and rising in the pressure vessel like a stormy sea, alternately leaving its core exposed or feeding water into the turbine-line – abetted by both junior and senior operators doing one wrong thing after another. Control was recovered more by luck than management. In questions of reactor safety, luck is much too fickle to count on.

It is necessary to add one more word, with particular reference to the safety of liquid-metal-cooled fast breeder reactors. In contrast to most thermal reactors, the liquid metal coolant of a fast breeder is under near-atmospheric pressure, putting much less strain on the reactor vessel and piping. In addition its thermal conductivity is high, so high that it is claimed that adequate cooling would be achieved even if the coolant circulation failed. Nonetheless, the coolant is flowing at very high speeds; even a slight blockage could impose sudden severe strains on pipework and internal structures. Such a blockage, with the resultant cooling impediment, led to the partial meltdown of the core of the Detroit Edison fast breeder reactor in 1966 (see, for instance, Novick, Bibliography). If the coolant should be lost, or even be allowed to reach boiling temperature, with formation of bubbles in the core, the consequence would almost certainly be very rapid meltdown. Needless to say, metallic sodium reacts explosively with water and indeed, at these temperatures, with many other substances, even including air; there is no possibility of emergency core cooling in a fast breeder. In addition, to all the problems associated with equipment malfunction and operator error, there must be added one last *caveat*. The fuel in a fast breeder, unlike that in a thermal reactor, might, during the course of a meltdown, collapse into a shape in which the concentrated fissile nuclei could produce a fast chain reaction: that is, a nuclear explosion.

Which is where we came in.

Running releases

Virtually any substance which spends much time in the company of neutrons will wind up radioactive. As a result, reactors tend to make their entire neighbourhood radioactive. So long as the components of this neighbourhood remain within the biological shield, all is well; but radioactivity finds a number of ways of escaping from the confines of reactors, however well buttoned-up.

Any radioactivity which emerges outside the biological shield in the course of routine reactor operation is called a 'running release'. The simplest kind of running release originates just inside the biological shield itself. In reactors with pre-stressed concrete pressure vessels which also serve as shielding, it is desirable to keep the concrete from being exposed directly to the heat of the core. Accordingly, a thin sheet of air is blown up the inside wall of the concrete, just outside the stainless steel liner which helps to keep the concrete gas-tight. The neutron flux in this neighbourhood bombards the various atoms present in the air. Some of them absorb neutrons and become radioactive, a process called 'neutron activation'; the most notable of the 'activation products' is a radioisotope of the inert gas argon. Some reactors are known to discharge hundreds of thousands of curies of argon 41 annually. Fortunately, however, argon 41 has a very short half-life, only some 1.8 hours; so this apparently enormous output decays to a very low activity before drifting from stacks down to ground level. Other atoms in air also become activated, but only in small amounts and/or for very short half-lives; the most important is nitrogen 16.

Moving closer, the next region of interest is that of the moderator itself. Impurities in the moderator – whether graphite or water – are susceptible to neutron activation. Graphite itself, if vaporised by high temperatures, can enter the coolant; and the coolant in most such reactors is carbon dioxide. Both categories of carbon may become radioactive carbon 14; note, though, that this requires for normal carbon 12 the absorption of not one but two neutrons, and is therefore relatively infrequent. (The same is true for the nitrogen 16 aforementioned.) Water molecules, containing both oxygen and hydrogen can be activated respectively to radioactive oxygen 18 and hydrogen 3 – also known as tritium; but once again the activation requires absorption of two neutrons, so the coolant does not, as a rule, become very active. The one coolant which responds with readiness to neutron activation is the liquid sodium coolant favoured for fast breeder reactors. It becomes sodium 24, so highly radioactive that it must be kept entirely within the biological shield.

The fuel cladding itself makes a contribution to the activity in the cooling circuit, as it suffers gradual corrosion by the hot coolant. Again it is primarily a question of impurities in the cladding – which has, of course, been made as little susceptible to neutron absorption as possible, for reasons of neutron economy. The worst offenders here are the impurities in the zircaloy cladding on water-cooled reactor fuel. Corrosion of this fuel is enhanced by the intimate contact with the liquid coolant at high pressure, and surface corrosion is quickly carried into the moving liquid.

Much more serious is the situation when the fuel cladding develops a leak: which cladding seems prone to do. As fuel undergoes fission, there accumulates within it a gradual build-up of gaseous fission products; we have already mentioned, for instance, xenon 135. The pressure of this 'fission gas' imposes an increasing strain on the cladding, in opposition – in most thermal reactors – to the external pressure from the coolant. If for any reason a fuel pin has a flaw in its cladding the fission gas will seek it out, and escape into the coolant. A more sizeable leak will also release the volatile fission products, among them the dangerous iodine 131 (see earlier in this chapter). In a Magnox reactor a major leak is called a 'burst can'; Magnox reactors have 'burst can detection gear' which are designed to sense the radioactivity in the coolant and scram the reactor automatically. Fuel which is leaking slowly may be difficult to locate; and, since to change it will disrupt the fuel cycle, with its planned effect on neutron flux in the core, leaking fuel is often left in the reactor until removed in the course of routine refuelling. Of course, each time a refuelling machine is coupled to a reactor vessel for on-load refuelling the machine acquires a share of the activity in the coolant; this activity must be discharged (and kept track of) when the machine is depressurised.

As a result of these combined effects it becomes necessary to decontaminate the cooling circuits of a reactor. Otherwise the unavoidable leakage of radioactivity through valve seals and other permeable points becomes a hazard to personnel. Decontamination is usually done routinely, by bleeding off a small portion of the coolant and replacing it with fresh uncontaminated coolant. In the case of carbon dioxide coolant the gas bled off – the 'off gas' – is passed over a variety of filters and delay stages. For instance, a particular kind of filter will trap almost all the volatile iodine in the coolant; such filters, as it happens, were the last line of defence which kept down the level of iodine 131 allowed to escape to the surrounding countryside during the fire in the Windscale reactor (see earlier in this chapter). Delay beds may, for instance, be activated charcoal; off-gas passed through such beds will have much of its radioactive content, especially that which is in the form of fine particles, adsorbed on to the charcoal, at least sufficiently long for the activity to decay to much lower levels.

Such measures are insufficient to reduce the level of

activity in coolant bled from some light-water reactors. Boiling-water reactors, which pass the primary coolant directly through turbines, are especially prone to problems of leakage of active coolant. One possible procedure in such a case is to provide storage tanks for coolant bled off. In such tanks the coolant activity can be allowed to decay for some months before it is released. Similar procedures are applied to drainage from floors, to sinks in which contamination is transferred to water, and to the laundry in which contaminated clothing is cleaned. In addition, the cooling ponds for storing irradiated fuel prior to shipment for reprocessing usually pick up activity from the exterior of the fuel elements and also from any leaking elements; so this cooling water, too, has to be dealt with. A common procedure is to cycle the water slowly through the ponds, continually diverting a small fraction, to be mixed with the much greater mass of water discharged from the turbine cooling system back into the river or coastal waters from which it has been abstracted for cooling purposes. (In this connection it is important to distinguish between the reactor coolant and the turbine coolant. The latter never enters the reactor; turbines, however driven, need to be cooled, and this coolant carries away the heat whether it be from fission, coal-burning, oil-burning or whatever. As it happens, the turbine coolant carries away a larger percentage of the input heat if this heat comes from a reactor, because of the low efficiency of the cycle; but waste heat, so-called 'thermal pollution', is by no means peculiar to nuclear power stations, and will not be discussed here.)

With regard to discharges of radioactivity, the regulatory bodies, the regulations they set and the way in which they carry out the necessary policing, vary markedly from country to country. The ranking body is the International Commission on Radiological Protection, ICRP, whose membership is made up of leading experts in radiobiology and related subjects. Their publications provide a baseline from which national authorities lay down suitable local criteria. Much controversy now surrounds both the basic ICRP guidelines and the national criteria; for further information consult the Bibliography. It is here worth mentioning, as an indication of the trend, that the basic ICRP guidelines on 'acceptable' radiation exposure (and therefore by implication on running releases) have shown a marked downward trend since the establishment of the ICRP in 1928. The more we learn about radiation the more cautious we are becoming. Familiarity with radiation is breeding not contempt but concern.

4

REACTOR-RELATED OPERATIONS: THE 'FUEL CYCLE'

Uranium mining

Uranium is found in nature as mineralisation in hard dense grey rock. Deposits occur in many parts of the world, notably Canada, Australia and Africa. But the deposit to achieve earliest prominence was in the Joachimsthal, in Germany. It was worked as an underground mine. By the 1930s it was noticed that Joachimsthal uranium miners had a tendency to develop lung cancer; the cause was traced to the radioactive gas radon, which is given off as a decay product of radium. The gas in the mine would be inhaled by a miner; in his lungs it would undergo another radioactive decay resulting in solid 'daughter products' which would remain in the lungs. These 'radon daughters', radioactive alpha emitters, would cause damage to the miner's lungs often leading in due course to lung cancer.

The surge of military interest in uranium began in 1940, and focussed particularly on deposits in the southwestern US. Over a period of years, many workings were active, both underground and open-pit. Uranium is extracted by grinding the mineralised ore to the consistency of sand and dissolving out the uranium. The result is a mixture of uranium oxides in a proportion that can be written U_3O_8, usually called 'yellowcake'. High-grade uranium ore usually contains 1 to 4 per cent U_3O_8. Such ore occurs in the Canadian North West Territories, and in the African Congo basin. Grades of nearer 0.4 per cent are now being worked in the US, southern Africa and elsewhere. Grades much lower than this are available but not yet being extensively worked. Besides the U_3O_8, there remains after extraction some 100 times its weight of sand, called 'tailings' – which also contains the radium; and there also remains, per ton of ore, nearly 900 gallons of liquid wastes, both chemically toxic and radioactive.

The military rush for uranium in the US led to the accumulation of vast piles of tailings; estimates range as high as 90 million tons. Only recently has it been discovered that these tailings have been used as a structural material in concrete for many buildings in at least nine different towns in the south-western US: buildings including homes, schools and hospitals, all of which are now vulnerable to the gradual build-up of radon daughters carried up from their radioactive foundations. The people living in these areas are now exposed to the hazards which – despite the well-documented case of Joachimsthal – have led to a marked increase in the rate of lung cancer among American uranium miners.

Only in late 1972 did the US Government finally agree to contribute funds to control the piles of dry tailings still blown freely by the wind across many inhabited areas of the south-western US. The liquid wastes, too, were cast away with little thought; so that downstream in the Colorado River basin, inhabitants have been at times exposed to up to three times the ICRP maximum permissible intake of radium – a bone-seeking radionuclide even more dangerous than strontium 90.

Uranium enrichment

The fissile uranium 235 nuclei in natural uranium are too dilute to support a chain reaction, as described earlier (see Chapter 1). For weapons, and for many types of reactor, it is desirable to increase the concentration of the 235 isotope in the uranium. This cannot be done chemically: in chemistry both isotopes are virtually identical. Only their minute difference in mass – 3 units in 235 – can be used as a basis for separation. This mass-difference slightly affects how fast the two isotopes, in a gaseous uranium compound, diffuse through a porous membrane; the effect became the basis for what are arguably the largest industrial establishments in the world, so-called 'gaseous diffusion plants'. There are three such plants in the US, including the vast Oak Ridge plant which enriched the uranium for the Hiroshima bomb; one in Britain, at Capenhurst; one in France; one in the USSR; and one in China.

The only uranium compound which is gaseous at room temperature is far from ideal. Called uranium hexafluoride – UF_6 – it is a viciously corrosive, poisonous, reactive gas, requiring very careful handling and high-quality metallurgy in the vessels through which it must travel.

The basis of a gaseous diffusion plant is very simple: a metal-walled cell, with a thin membrane of porous metal dividing it in two. Uranium hexafluoride, made chemically from 'yellowcake', is pumped into one chamber of this cell; under pressure it diffuses through the membrane, and the lighter 235 isotopes diffuse slightly more readily than the heavier 238 isotopes. The improvement is, however, so slight that the process must be repeated thousands of times. A cascade arrangement is set up: gas from the first chamber, slightly depleted of the 235, is piped back to earlier cells; gas from the second chamber, slightly enriched in 235, is piped onward to later cells. By this means, using thousands of pumps and condensers, it is possible to raise the proportion of 235 isotope in the uranium close to 100 per cent. (The uranium whose share of 235 nuclei has been reduced is called 'depleted' uranium; it is used for shielding and as fertile 'blanket' in breeder reactors.)

All seven of the world's gaseous diffusion plants were built under military auspices. Their electrical requirements are awesome; the Oak Ridge plant, in full operation, requires some 2000 megawatts of electricity, enough to power a sizeable city. (This electricity is provided largely by fossil-fuel power plants burning strip-mined coal, a nicely ironic touch.) Of course for non-military fissile material the level of enrichment required is much less; fuel for light-water reactors or AGRs need be only two or three per cent uranium 235.

The British gaseous diffusion plant at Capenhurst has been almost completely shut down since 1961; even the advent of the AGRs has not required more than the early stages of the plant.

Nonetheless, interest is burgeoning in an alternative approach to enrichment, which requires only one-tenth as much electricity per unit of 'separative work' – the imprecise designation always used in describing the enrichment process. This process once again passes uranium hexafluoride through a cascade, but this time the units of the cascade are small centrifuges. The gas is pumped along the axis of the centrifuge, and the heavier nuclei tend to drift towards the outer circumference, leaving the inner portion richer in uranium 235. The centrifuges are linked in cascade, and once again the degree of enrichment achieved depends on how many cascades are used. Experimental gas centrifuge plants are now being set up in Britain, alongside the gaseous diffusion plant at Capenhurst, and in Germany and the Netherlands (see Chapter 5).

Fuel fabrication

Fabrication of fuel elements for reactors is now a complex industrial process. The essential requirement is purity of the uranium, particularly freedom from unwanted neutron-absorbing impurities. Structural materials in fuel elements, including cladding and supports, have to function under extreme conditions of temperature, pressure and neutron bombardment. From uranium metal interest has shifted to less awkward formulations; the metal has a number of changes of crystal structure associated with temperature changes, whereas powdered uranium oxide or even uranium carbide can be compacted into ceramic pellets of stable dimensions and high melting point. For fast reactors combinations of uranium and plutonium oxides are now under test. Notice must also be taken of the need to reprocess fuel once it is irradiated; certain fuels are much more manageable than others. But undoubtedly the most pressing matter now on the agenda in fuel research is to discover what has been happening to certain types of fuel element used in PWRs; see Chapter 3 and Bibliography. The AEC rule-making hearings on 'the fuel cycle' began on 1 February 1973 – they should prove revealing.

Transport

One of the major advantages claimed for central electricity generation using a nuclear heat-source is the relatively small bulk and mass of fuel and waste that must be transported to and from the station. A fossil-fuel station will require so much transport of coal or oil that it is a marked advantage to situate the station near the supply. A nuclear power station, on the other hand, will require at most one or two shipments a week; furthermore, the shipments away from the station are far more massive and bulky than those to the station. This is, of course, because fresh fuel rods, only minimally radioactive, can be and are shipped in ordinary cases like any other cargo. But once irradiated, they must thenceforth be heavily shielded, so that a shipment of two tons of irradiated fuel requires a 50-ton steel shipping cask.

Fresh reactor fuel is shipped by rail, by road, by water, and by air, in increasing quantities every year, from the suppliers to the reactor-sites. Apart from the usual protections against low-level radioactivity around the shipping cases, the main technical consideration is that of guarding against intentional or accidental stacking of cases so close together that the group of fissile material can reach criticality. (But see Chapter 5.) Elaborate codes of practice are published to provide appropriate guidelines.

The shipping of irradiated fuel is something else again. The irradiated fuel must be handled remotely, at every stage, from loading at the station – usually from a cooling pond – to unloading at the reprocessing plant (see next page). For short journeys it usually travels in massive casks filled with water and vaned on the exterior to help dissipate the decay heat; for longer journeys, especially those by sea, the casks must be coupled to cooling circuitry.

An accident involving a shipment of irradiated fuel could release dangerous amounts of radioactivity: particularly if – as is now being mooted – the fuel is shipped after not 120 days, or even 90 days, in a cooling pond but (for reasons of economy – fuel is expensive) after as little as 30 days, when even short-lived isotopes will still be contributing significantly to the radioactivity. Irradiated fast breeder fuel will pose even greater problems. Containers must pass severe tests, typically a 30-minute fire after a 30-foot fall. But as the number of shipments increases, so does the chance of an accident.

Reprocessing

One feature distinguishes nuclear power technology from all others: the left-overs. Unlike the ash, say, from a coal-fired station, the used fuel from a nuclear power station contains both very valuable material and uniquely troublesome waste. Recall that the first large reactors were built expressly so that, under neutron bombardment, the uranium 238 in the fuel would be transmitted into plutonium 239. This plutonium had to be recovered, as did the unused uranium 235 which was still left after fission products had poisoned the chain reaction. The same requirement holds today; both plutonium and uranium are much too valuable to throw away. Nor must the remainder of the fuel, the fission products, be thrown away – not because of their value but because of their dangerous radioactivity. So the irradiated fuel from a reactor must be 'reprocessed'.

A reprocessing plant is a chemical plant – but no ordinary chemical plant. Because its raw material, irradiated fuel, is intensely radioactive, all the operations must be carried out by remote control, behind shielding. The process equipment must be highly reliable, and require a minimum of maintenance: once it has been contaminated by the radioactivity, any malfunction will necessitate months, or indeed years, of decontamination before it can be set right. Accordingly the process line uses a minimum of mechanical parts, and depends instead on gravity-flow and simple valves.

Different designs of fuel require different handling. The British reprocessing plant, at Windscale, was set up to handle fuel elements from the plutonium reactors and the Magnox reactors. Fuel elements enter in shipping casks, which are opened by remote control, while operators watch on closed-circuit television. Transferred to massive shielded transport cases, the fuel elements are lifted 15 storeys to the topmost floor of the plant, and fed into the first of a series of 'hot cells'. Operators viewing through yard-thick double windows filled with orange-tinted bromine solution pick up the elements by remote control and drop them on to a stripping machine which unzips the metal cladding as easily as peeling a banana. The contaminated cladding drops down a chute into the thick concrete bin which extends from ground level to the tenth storey of the plant, there to remain indefinitely. The bare fuel rod is chopped into short slices and dropped into a vat of acid, which dissolves it.

Zircaloy-clad fuel is treated the same way, except that entire fuel elements, up to nine inches in diameter, are simply chopped into slices without being stripped. The irradiated fuel is dissolved out of the cladding by acid. The zircaloy remains fall into a bin adjoining that for Magnox. By chemical means the acid solution is separated into three streams: one containing uranium, one containing plutonium, and one containing the fission products. The uranium and plutonium streams each pass through recovery plants and emerge again as solid compounds, ready to be returned for fabrication into new fuel elements (or weapons).

In the course of the process, gases – notably radioactive krypton 85 – and liquids from the hot cells accumulate; they and other dilute radioactive fluids from contaminated areas are discharged through stacks, or out to sea by means of a pipeline two miles long.

The fission-product stream is concentrated as much as feasible, to reduce its volume, after which it passes through a 5 cm pipe encased in a 4 m concrete conduit to another building nearby, in which are the waste storage tanks. To call these vessels 'storage tanks' is to do them less than justice. They are in fact elaborate refrigerators: double-walled stainless steel chambers, about the size of a small room, containing seven separate circuits of cooling pipes. Each tank is situated in a concrete cubicle, lined on the floor and up to head-level with stainless steel. There are at present nine tanks each of 70 m^3 capacity, and three of 150 m^3 capacity. The most recent tanks are still under construction, and accessible; but the tanks in use are permanently walled in behind thick concrete shielding, never to be seen again. If a tank in use should develop a flaw, its contents can be pumped into standby tanks kept for the purpose.

Similar facilities are in operation in several other parts of the world. The most famous is at Hanford, where the waste from the military plutonium production is stored in some 150 huge tanks, of which some have already begun to leak.

The problem of the high-activity waste from fuel reprocessing is probably the most daunting of all the problems posed by nuclear reactor operation. A significant proportion of the radioisotopes in the storage tanks are long-lived; their activity will not fall below dangerous levels for decades, or, in some cases, centuries and indeed millenia. The fission product radioactivity, once created, can never be destroyed; it must die away of its own accord, in its own time. Occasional suggestions refer to the possibility of transmuting long-lived waste to short-lived; but this would require more energy than the nuclear fuel itself could ever produce. It has been seriously proposed that high-level waste might be fired from the earth by rocket to the sun; but once again the cost would be – excuse the expression – astronomical: requiring eventually several launches per week, with the ever-present danger of a rocket failure dumping the waste back to earth. Other equally futuristic proposals include the notion of dumping such waste in casks on to the ocean bottom, where geological movement might gradually swallow them into the earth's interior. But a realistic view is that we are stuck with however much high-activity waste we create – and so are our children, and our children's children for centuries hence.

The waste cannot be simply left to itself. Under the action of the hot acid, tanks will corrode and must be replaced. The cooling must be maintained, lest the liquid boil and burst the tank. Accordingly, attempts are now underway to find a way to solidify the waste and at least simplify the storage problem. It has been found possible to 'glassify' the waste: to evaporate it and melt the solid into a – hopefully – impermeable glass brick, which can then be stored, perhaps, underground. But the annual waste from a single 1000-MWe power reactor would still require about 15 cylinders some 30 cm in diameter and 3 m long: and these cylinders would be both hot and highly radioactive. Hopes are being pinned on the possibility of storing such cylinders in salt caverns underground. It is believed that the salt, heated to melting by the radioactivity, would provide protection against contact with ground water, and adjust itself into snug heat-conductive packing around the cylinders. But one attempt at least, at Lyons, Kansas, came to grief when the salt beds proved to have unexpected fissures. Canada has flatly announced that her high-level waste is to be stored above ground in tanks until a fully proven technique has been established for any alternative method.

It is worth noting in passing that the nuclear industry refers, as a matter of course, to 'waste management'. It looks like a career with a future – a long future.

5

THE REACTOR BUSINESS

Nuclear Byzantium

The involuted ramifications of the reactor business are entirely in keeping with its esoteric technology and its military origins. Factors of importance include: the roles of government, the military and private industry; the finances of reactor research and development, construction, operation and servicing; the promotion of nuclear power *vis-à-vis* its policing; the comparative economics of energy, nuclear and non-nuclear; insurance – a very telling factor, as we shall see; international arrangements, military and non-military, and their administration; and every conceivable cross-link between these vast subject areas. Clearly no brief exegesis can do the situation justice. But a few indicative comments may make it easier to disentangle the reports, commentaries, analyses, forecasts and plain old down-home propaganda with which every reactor-watcher must contend.

The USAEC

The military importance of nuclear energy has meant that virtually every country with a significant interest in nuclear development exerts this interest through the agency of a government body. Some of these bodies are listed in Appendix C. Of these the largest and by far the most important is the United States Atomic Energy Commission – the AEC. Governmental nuclear energy organisations in other countries are more fragmented, have a much narrower jurisdiction, are much less dedicated to the care, feeding and propagation of reactors, and have much smaller budgets.

The AEC has had from its inception a dubious dual responsibility. It must simultaneously promote the development of reactor-technology, and set and police all the relevant standards, including those concerning reactor safety and running releases of radioactivity. The obvious conflict of interest has been repeatedly challenged, but it seems as far as ever from resolution. No short description could do the AEC justice. Its own publications give one side of the story. There are now several others, which provide valuable counterweight, particularly those by Rapoport, Metzger and Lewis (see Bibliography).

The IAEA

In a famous speech in 1953 President Eisenhower called for the establishment of an international United Nations body to administer and coordinate a collective interest in nuclear energy for the good of humanity; rather than, as was the case at the time, exclusively for the detonation of ever-larger hydrogen bombs by the US and the USSR. The result was the International Atomic Energy Agency, created in 1956 at the first international conference on 'Atoms for Peace'. The proceedings of the subsequent Atoms for Peace conferences – thus far four in number – constitute a valuable compendium of shared nuclear experience. Apart from other less overpowering, more specialised, gatherings, the activities of the IAEA seem notable mainly for out-AECing the AEC in enthusiasm for fission energy. Sigvard Eklund, current Director-General of the IAEA, went so far as to tell the United Nations Conference on the Human Environment at Stockholm in June 1972 that within 30 years mankind would be starting up new fission reactors at the rate of one 1000-MWe reactor per day.

The NEA

Another international body of some importance is what was until 1972 called the European Nuclear Energy Agency. Since Japan has recently become a member it is now simply the Nuclear Energy Agency, one of the specialised agencies of the Organisation for Economic Cooperation and Development, the OECD. It, too, holds conferences and symposia, and organises exchanges of expertise and technology. Of perhaps more general interest, it also arranges deep-ocean dumping of low-level radioactive waste from member states, in an unidentified location in the North Atlantic.

The nuclear moneymakers

There are many other collective groups, of varying scale and effectiveness – some are listed in Appendix C. But their importance is outranked by the growing importance, both economic and policy-making, of some private and semi-private industrial concerns. Private industry has shown considerable interest in nuclear technology since the early days, because of the promise of lucrative defence-related contracts; because of the threat posed to their own market structure by the intervention of the government bodies; because of forthright arm-twisting by some government bodies themselves; and because, in many instances, it appeared that possible profits were likely to accrue to the private corporations while the government side – that is, ultimately, the taxpayer – carried the economic risks. These risks related to research and development, to financing of construction and operation and – most especially – to insurance against catastrophe. It soon became clear that private corporations would be able to use reactor data acquired in a military context for civil purposes; would receive backing amounting effectively to subsidy for their own efforts; would have the support of enrichment and reprocessing facilities established for military purposes; and would not be required to provide more than a small fraction of the necessary insurance coverage for their undertakings. On these terms, private industry leapt aboard the nuclear bandwagon with mounting enthusiasm. The most notable successes have since been notched up by the three American multinational giants Westinghouse (makers of PWRs) and General Electric (makers of BWRs), who between them have a good start on girdling the globe with light-water reactors, and Gulf General Atomic, who have recently bought a large share of European high-temperature reactor expertise.

It would be uncharitable to do more than note in passing how the spirit of 'free enterprise' appears to thrive when the corporation takes the returns and the government takes the chances.

Nuclear insurance

Energy economics is becoming a major discipline in its own right. One of its most urgently needed areas of study is that of the comparative costs (and benefits) of nuclear as against non-nuclear energy sources. Is it too much to hope that any such study would accept as a third alternative the more fundamental possibility of simply using less energy? As it is, the ascertaining of true costs, capital and running, taking account of research and development, reasonable amortisation of plant, subsidised services, and countless other niceties, is a challenge crying to be met.

The most blatant example of dishonest financing of nuclear technology occurs in the field of insurance. In the US in the mid-1950s it became apparent, especially after publication of WASH-740, that electrical utilities were shying noticeably away from the 'promise of cheap, inexhaustible power' ostensibly theirs for the taking. The problem was simply that no insurance company, indeed not even a huge consortium of insurance companies, could be persuaded to provide coverage against the possibility of a major reactor accident. Although insurance could be obtained for almost any other conceivable eventuality, the mind-numbing consequences of a massive release of radioactivity froze the insurance companies in their tracks.

Accordingly, two members of the Joint Congressional Committee on Atomic Energy (JCAE), Price and Anderson by name, drafted, and in 1957 won Congressional backing for, the Act which now bears their names. The Price–Anderson Act specifies, in effect, that reactor operators shall chase up as much coverage as they can persuade private insurers to offer – even now only some $66 million. To this the US Government adds another $500 million. Beyond this – remember that WASH-740 foresaw property damage alone reaching $7000 million – it's every man for himself. Furthermore, if no claims have to be covered, even the grossly inadequate premiums paid by the operators are eventually refunded – which must make other industries, not to mention ordinary householders, more than somewhat envious. In 1965, two years before the Price–Anderson Act was due for renewal, the renewal was hustled through, lest later objections prove an embarrassment to the suddenly uncovered nuclear industry. Accordingly, 1977 will be an interesting year in the US nuclear business.

Meanwhile, a similar pantomime was taking place in Europe, both nationally and internationally. In Britain, for instance, according to the Nuclear Installations Acts of 1965, a reactor operator need provide coverage for only £5 million liability; the government adds another £43 million – and that's the lot, which makes even Price–Anderson look generous. Furthermore, there have been efforts since before 1960 to reach international agreement on insurance against nuclear hazards; the vagrant habits of radioactive clouds are all too well documented. But the efforts have been thus far quite in vain. Draft documents have been circulated several times, but not ratified.

Nuclear economics

Despite the feather-bedding, the nuclear industry has long made great play with the economies to be gained by nuclear generation of electricity. In the early days – the late 1950s in Britain, the early 1960s in the US – it was not unusual to be told that nuclear power would be so cheap that it would totally wipe out the coal industry. Within a decade the nuclear proponents have done a somersault. Indeed, in the past two years the warning has gone out that supplies of uranium are growing scarce; that the fast breeders are arriving in the nick of time to generate new fissile fuel, since natural supplies cannot otherwise last out the century. Neither extreme position bears close examination.

The early euphoria has, of course, been long since discredited. Nuclear fuel may have its advantages, but its manufacture is a lot more complicated than breaking coal. Nuclear plant, too, involves capitalisation perhaps five times as high per kilowatt of output as does fossil-fuel plant. When the nuclear plant involves new and otherwise untried technology, and must be debugged as it is built, the costs have a tendency to spiral skyward, and construction schedules seem the stuff of fantasy, even without external hindrance from neighbourhood doubters.

Nonetheless, having displayed their inability to forecast the current situation accurately, the forecasters are moving enthusiastically onwards, and insisting that energy 'demands'/'requirements'/'needs' – vigorously encouraged by advertising and promotional rates – can only be met by a crash programme to construct commercial fast breeder reactors. Otherwise the supplies of uranium presently mined, costing (in the US) only some $6 to $8 per pound of yellowcake, will soon be exhausted, and more expensive uranium must be mined. This argument ought to be examined minutely. The cost of fuel – that is, of ore – for a reactor is a much smaller proportion of its total cost than is the cost of fuel for a fossil-fuelled station. Accordingly, an increase in the cost of uranium, even an increase by a factor of 10 or more, will produce only a trivial increase in the cost of the electricity generated. The supplies of cheap uranium, at least those in proven reserves, are indeed not likely to last very long; but moving to more expensive uranium (that is, to lower-grade ores requiring more processing) extends the available reserves by centuries. That being the case, the urgency of fast-breeder development looks more than somewhat specious, especially since – as usual – the taxpayers are taking all the risks, both financial and physical.

Cross-examining your friendly neighbourhood reactor

One way and another, the international nuclear industry, though pampered by governments, is more and more facing informed dissent, and problems are less and less readily possible to conceal or deny. In the early days a reactor was a military installation: it could be set up anywhere, and protests were not only unpatriotic but futile. This is no longer so. Since the historic opposition to the plan for a power reactor on the California coast at Bodega Bay succeeded, nearly 10 years ago, in stopping its construction, reactor-siting has become a matter not only of economics, geology and hydrology, but of politics; and the opponents may also have a good deal to say about the geology and hydrology. In the US and on the European continent there has in the past five years been an unending series of confrontations over planned construction of reactors – although it is true that there has been, thus far, virtually no opposition to reactor-siting proposals in the UK. But signs are increasingly evident that the international fellowship of reactor-adherents will soon be matched by an international coalition of opponents.

The whole life-cycle of a reactor is now being questioned. The cross-examination includes at least the following questions:

1. Why is it to be built? If for power, is nuclear power clearly the best choice? Is the power itself clearly needed?

2. Why is it to be built at that location? Has note been taken of possible seismic hazards? (If a reactor were to encounter an earthquake, even a pre-stressed concrete containment might not survive intact.) What about tornados? Hurricanes? Floods? What about cooling? What about local ecology? Aesthetics? Should it be so close to centres of population? In the event of an air crash, might it be underneath? Should it be underground?

3. Who is to pay for it and how? (This is never easy to find out, but even in the most aggressively 'free enterprise' context it is a near-certainty that the taxpayers' money will be in there somewhere – and not just via his electrical bills.)

4. What are the benefits, and who gets them? (Orders for plant and machinery, local employment on construction, and similar factors will be cited, and will be significant. But remember the answers to question 3.)

5. Who will run the facility? (By this is meant not 'what board of directors?' – although that is relevant – but 'what qualified staff, and how qualified?' Does a junior technician know what a mistake on his part might lead to? Does he care? Do his superiors?)

6. What safety features does the plant embody? Do they work? Who says so – merely a computer?

7. What contingency plans and emergency procedures are envisioned? Will they be adequate to limit the consequences of an accident? Or will they just make the consequences better documented?

8. What insurance cover will the plant carry? How will it be financed? By whom?

9. What running releases of radioactivity will take place during normal plant operation? Who will measure them? What will be their effects? Are they necessary? That is, would the extra cost of not releasing radioactivity make the nuclear option less economically enticing?

10. What services will the plant require? In particular what fuel shipments will be involved? By what means? Along what routes?

11. What will become of the plant at the end of its

useful life? (Nowhere in the world has anyone thus far dismantled a large power reactor which has been running for years. The problem of 'decommissioning' even small reactors is considerable. The fuel and coolant can be removed, but nothing can be done about the radioactivity of the core materials. It seems probable that a site on which a large reactor has operated will have to be 'dedicated in perpetuity' – that is, left with the hulk of the reactor, possibly entombed in concrete, as an everlasting monument to, say, 30 years of electrical output.)

12. What security can be offered, at every stage, against damage which is not accidental but wilful? – that is, against sabotage?

The last question

The last question is probably the single most frightening question to ask about reactors. No convincing answer has yet been forthcoming – nor is one likely. Any facility which incorporates an inventory of radioactivity vast enough to make a sizeable area of the planet uninhabitable for the foreseeable future is terrifyingly vulnerable to the single-mindedness of contemporary fanaticism. In this category are numbered high-level radioactive waste storage facilities, fuel-reprocessing plants, and large power reactors. The more of these there are, the more difficult it will be to prevent a team of suicidal diehards from breaching the protective systems. Remember that these are not just passive vaults: they contain in themselves stored energy that will be on the side of the saboteurs when it comes to releasing radioactivity. A well-placed explosive charge could knock out both the controls and the emergency systems – and the waste tank or reactor itself would take over from there.

More subtle still are the problems posed by large-scale fast breeder technology. Particularly if fast breeders proliferate, nuclear power will soon lead to a 'plutonium economy', with growing inventories of this dangerous substance not only accumulating in many localities, but travelling continually from one location to another. Despite military secrecy, it is now well-known that a team of relatively untrained technicians could, with a single well-equipped workshop, process plutonium reactor-fuel so as to make an unpredictable but fully convincing nuclear explosive. All the necessary physics is in the open literature. There are already a number of hair-raising case histories of plutonium allowed to go astray through pure ineptitude in shipping (see Geesaman, in Stockholm Conference Eco, Bibliography); the possibility of hijacking of plutonium is all too real. Indeed, a similar black market might, for the same reasons, develop also for the highly-enriched uranium fuel used in high-temperature reactors. Nuclear industry executives, when challenged on this point, have a stock answer. They insist that of course they are aware of the situation, and have taken appropriate steps – and no, they cannot say what the steps are, because that would make them less effective.

It all sounds too much like whistling in the dark. But unless we change our attitudes towards energy, we shall have to rely increasingly on nuclear reactors. Next time, before you turn on the light, think for a moment. Has the energy you are going to use come from fissile nuclei? And can you hear whistling?

APPENDIX A: NUCLEAR JARGON

ACTIVATION: absorption of neutrons to make a substance radioactive.

AEC, USAEC: US Atomic Energy Commission.

AGR: advanced gas-cooled reactor.

ALPHA PARTICLE: high-energy helium nucleus (two protons, two neutrons) emitted by some heavy radioactive nuclei.

BETA PARTICLE: high-energy electron emitted by radioactive nuclei.

BORON: powerful absorber of neutrons used – usually in alloy steel – for reactor control rods, etc.

BURN-UP: cumulative output of heat from core directly correlated with cumulative build-up of fission products.

BWR: boiling-water reactor.

CADMIUM: powerful absorber of neutrons used for reactor control rods, etc.

CENTRIFUGE: see GAS CENTRIFUGE.

CHINA SYNDROME: consequence of core melt-down: molten mass of intensely radioactive material plummets through vessel and containment and into the earth in the direction of China (unless the reactor is, in say, Japan).

CLADDING: sometimes just CLAD (as noun): metal sheath (Magnox, zircaloy, stainless steel, ceramic) within which reactor fuel is hermetically sealed.

CONTROL ROD: rod of neutron-absorbing metal inserted into reactor core to soak up neutrons and shut off or reduce rate of fission reaction.

CONVERSION RATIO: number of fertile nuclei converted to fissile, compared to number of fissile nuclei lost undergoing fission.

COOLANT: liquid (water, molten metal) or gas (carbon dioxide, helium) pumped through reactor core to remove heat generated in the core.

COOLING POND: deep tank of water into which irradiated fuel is discharged upon its removal from a reactor, there to remain until shipped for reprocessing.

CORE: the region of a reactor containing fuel (and moderator, if any) within which the fission reaction is occurring.

CRITICAL: refers to a chain reaction in which the total number of neutrons in one 'generation' of a chain reaction is the same as the total number of neutrons in the next 'generation' of the chain; that is, a system in which the neutron density is neither increasing nor decreasing.

CRITICALITY: the state of being CRITICAL.

CRITICALITY ACCIDENT: inadvertent accumulation of fissile material into a critical assembly, accompanied by outburst of neutrons and gamma radiation.

DAUGHTER: as in DAUGHTER PRODUCT, the substance into which a radioactive nucleus transforms itself by radioactive decay.

DECAY: radioactive transformation.

DECAY HEAT: heat in core of reactor generated not by fission but by radioactive decay of fission products.

DEUTERIUM: 'heavy hydrogen' – its nucleus consists of one proton plus one neutron.

DEUTERIUM OXIDE: water in which the hydrogen atoms are heavy hydrogen.

DEUTERON: nucleus of heavy hydrogen.

DIVERSION: euphemism for theft, as applied to 'special nuclear material'.

ECCS: emergency core cooling system; safety device on water-cooled reactors intended to supply extra cooling water to remove decay heat from core in the event of a loss-of-coolant accident (LOCA).

ENRICHED: as in ENRICHED URANIUM, uranium in which the proportion of uranium 235 is higher than 0.7 per cent.

ENRICHMENT: process of making enriched uranium.

FAST: of neutron; high energy, as direct from fission.

FAST BREEDER: reactor designed to have conversion ratio greater than 1, using unmoderated fast neutrons.

FERTILE: of material like uranium 238 or thorium 232, which can be transformed by neutron absorption into fissile material.

FISSILE: capable of undergoing spontaneous fission, with production of free neutrons.

FISSION: rupture of a nucleus into two lighter fragments (FISSION PRODUCTS) plus free neutrons – either spontaneously or as a consequence of absorption of a neutron.

FLUX: of neutrons, moving cloud of particles, particularly in reactor core.

FUEL: material (such as natural or enriched uranium or uranium dioxide) containing fissile nuclei, fabricated into suitable form for use in reactor.

FUEL CHARGE: the total complement of fuel contained in a reactor core.

FUEL ELEMENT: unit of fuel for reactor.

FUEL PIN: in complex fuel element, a rod or tube containing fuel.

GAMMA RAY: high-energy electromagnetic radiation emitted by nucleus – very penetrating.

GAS CENTRIFUGE: uranium enrichment device by which heavier uranium 238 nuclei are slightly separated from lighter uranium 235 nuclei by centrifuging of uranium hexafluoride gas; full-scale plant uses many thousands of centrifuges in cascade.

GASEOUS DIFFUSION: uranium enrichment process utilising slight difference in rate of diffusion of uranium 235 and 238 hexafluoride molecules through porous metallic membrane; full-scale plant uses many thousands of diffusion cells in cascade.

HALF-LIFE: period of time within which half the nuclei in a sample of radioactive material undergo decay; characteristic constant for each particular species of nucleus.

HEAT EXCHANGER: boiler, in which hot coolant from reactor core raises steam to drive turbo-generators.

HEAVY HYDROGEN: see DEUTERIUM.

HEAVY WATER: see DEUTERIUM OXIDE.

HTGR: high-temperature gas-cooled reactor.

ISOTOPE: of an element, having the same number of protons in its nucleus as all other varieties of the element, but having a particular number of neutrons which may be different from that of other varieties of the element.

LFMBR: liquid metal fast breeder reactor.

LOCA: loss-of-coolant accident.

MAGNOX: magnesium alloy used in cladding for British gas-cooled reactors; generic term for reactors using such fuel.

MELTDOWN: of reactor fuel or core – consequence of uncontrolled temperature rise as result of malfunction, such that cladding, fuel and possibly support structure melt and collapse.

MODERATOR: material of low atomic weight (light water, heavy water, graphite) used in reactor core to slow down fast neutrons to increase probability of their absorption in uranium 235 to cause fission.

NEUTRON: uncharged particle, constituent of nucleus – ejected at high energy during fission, capable of being absorbed in another nucleus and bringing about further fission or radioactive behaviour.

PLUTONIUM: heavy artificial metal, made by neutron bombardment of uranium; fissile, highly reactive chemically, extremely toxic, both chemically and radiologically.

PRICE-ANDERSON ACT: in the US, Act of Congress limiting the insurance liability of reactor operators in the event of an accident.

PWR: pressurised water reactor.

RADIATION, NUCLEAR: neutrons, alpha or beta particles or gamma rays which radiate out from a radioactive substance.

RADIOACTIVITY: behaviour of substance in which nuclei are undergoing transformations and emitting radiation.

RADIOISOTOPE: radioactive isotope.

RADIONUCLIDE: radioactive nuclide.

REACTIVITY: measure of ability of assembly of fissile material to support sustained chain reaction.

RECHARGING: see REFUELLING.

REFLECTOR: of neutrons, material around a reactor core to reflect neutrons back into the reaction region.

REFUELLING: replacement of fuel when it has built up enough fission products to interfere with reactivity.

REM: Roentgen equivalent man; unit of radiation exposure, compensated to allow for extra biological damage by alpha particles or fast neutrons.

REPROCESSING: mechanical and chemical treatment of irradiated fuel to remove fission products and recover fissile material.

RUNAWAY: accidentally uncontrolled chain reaction.

RUNNING RELEASE: planned emission of radioactive material to the outside air or water.

SAFEGUARDS: term applied to keeping track of special nuclear material to prevent diversion.

SHIELDING: wall of material (concrete, lead, water) surrounding source of radiation, to reduce its intensity.

SNM: SPECIAL NUCLEAR MATERIAL: fissile material potentially usable in nuclear weapons.

START-UP: gradual withdrawal of control rods and accompanying procedures before a reactor goes critical.

SUB-CRITICAL: insufficiently supplied with neutrons to sustain a self-propagating chain reaction.

THERMAL: of neutron: low energy, sharing common heat-energy of surroundings; of reactor: functioning with chain reaction propagated with thermal neutrons.

TRITIUM: hydrogen 3 – nucleus contains one proton plus two neutrons; radioactive.

URANIUM: heaviest natural element, dark grey metal.

VESSEL (REACTOR): container (welded steel, prestressed concrete) enclosing reaction region of reactor; core, moderator, coolant, control rods, etc.

XENON POISONING: accumulation of neutron-hungry fission product xenon 135, reducing reactivity of reactor and necessitating refuelling.

APPENDIX B: BIBLIOGRAPHY

For those wishing to learn more about nuclear reactors, their attributes and implications, the following sources are variously useful.

The US Atomic Energy Commission publishes an enormous range of material, from the unintelligible to the trivial, with much of value in between. It is obtainable from the Superintendent of Documents, USAEC, Washington D.C., 20545, USA. The AEC's series of popular booklets 'Understanding the Atom' are free and informative, provided you read scrupulously between the lines. The sweetness and light is like an overdose of whipped cream.

The International Atomic Energy Agency, like the USAEC, publishes a wide range of material. For hard data the shelves filled with the Proceedings of the four Conferences on the Peaceful Uses of Atomic Energy are invaluable, albeit technical. The IAEA Directory is a – thus far – eight-volume compilation of data on all the world's reactors; the paperback *Power and Research Reactors in Member States* is a handy compendium. Their best popular publication is probably *Nuclear Power and the Environment*, not quite so euphoric as most industry propaganda. The IAEA also publishes a monthly Bulletin. IAEA publications are obtainable from the IAEA, Kärntner Ring 11, P.O. Box 590, A-1011, Vienna, Austria, or from government publications outlets in other countries.

Two other official bodies whose publications should be noted are the United Nations Scientific Committee on the Effects of Atomic Radiation, who have just recently published their latest survey on *Ionising Radiation*; and the International Commission on Radiological Protection, whose Publication 9 lays down the bases for most national restrictions on radioactivity.

The United Kingdom Atomic Energy Authority publishes, among many other items, the concise directory *Reactors UK* giving technical data on all British reactors. Their monthly bulletin *Atom* is also informative. Other national nuclear authorities should also be noted as sources for appropriate information.

The Nuclear Energy Agency (formerly the European Nuclear Energy Agency), among its other publications, recently presented an excellent report on *Radioactive Waste Management in Western Europe*, obtainable from the NEA, Boulevard Suchet, Paris 16, France.

Other official sources of interest are the various US Congressional Hearings, far too numerous to list, and, in Britain, certain of the sittings of the Select Committee on Science and Technology. Background papers for the US Congress – again too numerous to list – are also worthy of note; they are obtainable from the Library of Congress, Washington, D.C. Ask whether the particular topic in which you are interested has been covered.

The most concentrated compilation of basic nuclear

physics and related background information is undoubtedly *Sourcebook on Atomic Energy*, by Samuel Glasstone (Van Nostrand Reinhold, 1968): a one-volume encyclopaedia, clear without avoiding technicalities; it does, however, omit most of the less commendable side of the story, as might be expected of a long-time USAEC consultant.

Britain's Central Electricity Generating Board offers, in *Modern Power Station Practice: Nuclear Power Generation*, a superbly detailed and thorough discussion of how to design, build and operate your very own reactors. Not for the complete layman, but accessible; and honest about problems, at least on the engineering side.

The Careless Atom, by Sheldon Novick (Delta), is readable, purposeful and businesslike, one of the first 'popular' books on reactor problems and still among the best.

Perils of the Peaceful Atom, by Richard Curtis and Elizabeth Hogan (Gollancz, 1970), is one of the best-known and least satisfactory discussions of reactor problems; unrelentingly shrill and not overly scrupulous about accuracy.

Poisoned Power (Chatto & Windus) is an unattractive title, but anyone who knows how the AEC has treated its distinguished authors, John Gofman and Arthur Tamplin, will excuse the title. The book is an angry – but accurate – challenge to the AEC from within its own most qualified ranks – and has already won some tightening of standards. Highly recommended.

The Great American Bomb Machine, by Roger Rapoport (Earth Island, 1973), is a brilliant dissection of the unpleasant innards of the AEC; emphasis is on nuclear weapons, but the implications spill over copiously into nuclear power. Highly recommended.

The Atomic Establishment, by Peter Metzger (Simon & Schuster) is a scathing historical critique of the AEC in action, and of its cosy relationship with the Congressional Joint Committee on Atomic Energy, the watchdog that 'did nothing in the night-time'. Highly recommended.

Nuclear Power, by W. G. Jensen (Foulis), helps to place the role of nuclear energy in its economic and political context, giving valuable background to the European situation. Jensen is, however, neither physicist nor biologist, and the economics is on a traditional and uncomfortably narrow foundation.

The Nuclear Power Rebellion: Citizens vs. The Atomic Industrial Establishment by Richard Lewis (Viking) as an inside look at the many battle fronts upon which the American nuclear industry and its critics have clashed. The author, as editor of the *Bulletin of the Atomic Scientists*, has had a unique opportunity to observe the confrontation and indeed to further it.

Periodicals which maintain a wary lookout on nuclear developments include *Environment*, 438 N. Skinker Boulevard, St. Louis, No. 63130, USA, published by the St. Louis Committee for Environmental Information; *Bulletin of the Atomic Scientists*, 1020–24 E. 58th Street, Chicago, Ill. 60637, USA; *New Scientist*, 128 Long Acre, London WC2E 9QH, England, *Your Environment*, 10 Chesham Road, Amersham HP6 5ES, Bucks, England; and *The Ecologist*, Molesworth Street, Wadebridge, Cornwall, England; newspapers, particularly the *Guardian*, the *Observer*, and the *Financial Times*, all in Britain, and the *New York Times*, also bear watching.

Low-Level Radiation, by Ernest Sternglass (Earth Island, 1973), is one of the most controversial books on possible radiation hazards, variously attacked and defended from all sides. If Sternglass's thesis is even partially sound, we are already in serious trouble.

The testimony of the Union of Concerned Scientists before the AEC hearing on emergency core-cooling is, in part, abstruse and technical; but its second chapter, on the possible consequences of a nuclear accident, is all too vivid. Contact the UCS at P.O. Box 289, M.I.T. Branch Station, Cambridge, Mass. 02139, USA. They need your support.

The *Stockholm Conference Eco*, Vol. II, published August-September 1972, is an eleven-issue report centring on the AEC hearings about ECCS, but commenting widely on other aspects of nuclear power-production and its problems. Issue 11 includes both the Friends of the Earth testimony to the Joint Committee on Atomic Energy about fast breeder reactors, and excerpts from Donald Geesaman's testimony to California state hearings concerning the 'diversion' of plutonium. Some sets of *Eco* may still be obtainable from Friends of the Earth, 620 C Street, S.E., Washington D.C. 20003, USA.

George L. Weil, who started up the first reactor in 1942, has published privately a concise, wrathful, small book entitled *Nuclear Energy: Promises, Promises*, delineating the main problems faced – or ignored – by the American nuclear industry. It is obtainable from him at 1101 17th Street, N.W., Washington D.C. 20036, price $1.00.

The two periodicals which should be known to all reactor-watchers are *Nucleonics Week* and *Nuclear Engineering International. Nucleonics Week* is an international weekly newsletter published by McGraw-Hill, devoid of advertising and often sufficiently outspoken to bring aggrieved looks from the industry. But its subscription price is some $320 per year . . . check your library. *Nuclear Engineering International* is a glossy monthly published by IPC offering more in-depth material on nuclear developments and problems.

APPENDIX C: NUCLEAR ORGANISATIONS, PRO AND CON

International Atomic Energy Agency, Kärtner Ring 11, A-1011, Vienna, Austria.

Nuclear Energy Agency, Boulevard Suchet, Paris 16, France.

United States Atomic Energy Commission, Washington, D.C. 20545, USA.

United Kingdom Atomic Energy Authority, 11 Charles II Street, London, S.W.1, England.

United Nations Scientific Committee on the Effects of Atomic Radiation, United Nations, New York, N.Y., USA.

Atomic Energy of Canada Ltd., Congrill Buildings, 275 Slater Street, Ottawa K1A O54.

(For other national nuclear agencies try embassies for details.)

Union of Concerned Scientists, P.O. Box 289, M.I.T. Branch Station, Cabridmge, Mass. 02139, USA.

Consolidated National Intervenors, 153 E. Street S.E., Washington, D.C. 20003, USA.

Friends of the Earth, 620 C Street S.E., Washington, D.C. 20003, USA.

Friends of the Earth Ltd., 9 Poland Street, London W1V 3DG, England.

(Other national Friends of the Earth affiliates can be reached via the above addresses.)

Environmental Action, Room 731, 1346 Connecticut Avenue, Washington, D.C. 20036, USA.

Watch on the AEC, Citizen's Energy Council to Stop Environmental Pollution, 113 2nd Street, N.E., Washington, D.C. 20002, USA.

(National and local opposition groups can usually be reached through the appropriate affiliate of Friends of the Earth.)

Friends of the Earth and Earth Island

The present environmental crisis is now being recognised as an ultimate and perhaps not very remote threat to man's, and even 'our' planet's, survival. We are finally becoming aware that the land, sea and air on which we depend cannot tolerate the sort of abuse to which we subject them.

Action is required now to develop a life-style based on viable ecological principles, related to the earth's capacity to support the human and other species. But it is one thing to lament (and the capacity for lamentation is one which we humans have developed to a high degree through our history) the tragedy of dying rivers and the curse of industrial dereliction, and quite another to pick an issue, fight the case, and win.

Friends of the Earth Limited is Britain's leading environmental action group. In various translations it today exists in many countries across the world: an international, non-profit-making organisation, with a reputation for political flair and quick, imaginative thinking; a reputation for professionalism, and success.

FOE is light on its feet, academic in its research work, thorough in its strategy, and highly effective in action. Energy is directed into specific campaigns, at both national and local level. More than fifty affiliated FOE groups are now established throughout Britain, dependent on FOE, London, for information, advice, contacts and encouragement.

Campaigns to date (FOE was incorporated in London as recently as 1970) are those against wasteful or excessive packaging (our Government has *promised* us action); the endangered species campaign, aimed at forcing through legislation to ban the importation of the furs, feathers and other by-products of more than sixty animal species threatened with extinction; and the fight to save our National Parks – none is so threatened as Snowdonia – together with other areas of outstanding natural beauty under attack from certain mining companies in their (Government-approved) search for low-grade mineral and other deposits.

Further campaigns at the planning stage are those to tackle Energy (i.e. its waste); Transport; and Sewage. FOE fights its battles with the help of more than 3,000 members. It survives largely on their financial support. Join us, and beat them!

Friends of the Earth was founded in 1968 by America's leading, longest-serving environmentalist, David Brower. Goups now exist in the United States, United Kingdom, Australia, France, Germany, Holland, Italy, New Zealand, Norway, Sweden, Switzerland, and Yugoslavia.

Earth Island Limited is Britain's leading specialist environmental publishing house, linked with FOE, but like that action group, quite autonomous. Earth Island publishes only those books that *need* to be published. Environmental problems increase by the day. And as the realisation grows that all things are related in delicately-structured eco-systems, so the spectrum of what is properly 'environmental' literature expands. We shall nevertheless go on believing that most books would be best left as trees. . . .

INDEX

Non-technical index: for a Glossary of the most important technical terms, see Appendix A